Douglas *Skywarrior*

by René Francillon
with Edward H. Heinemann

Although this A3D-1 was photographed at the El Sugundo Division on April 14, 1956, more than thirty years ago, the aircraft's unusually clean lines ensure that it would still look modern if parked next to current generation business jets or light transports. The aircraft is finished in the new non-specular light gull gray and glossy insignia white scheme as specified in MIL-C-18263 (Aer).

INC.

Arlington, Texas

PUBLISHED BY

 INC.

P.O. Box 120127
Arlington, Texas 76012
Ph: (214) 647-1105

COVER PHOTOS
Top: ERA-3B, BuNo 144838, of VAQ-34.
Bruce Trombecky

Lower Right: ERA-3B, BuNo 144825, of PMTC.
Carl Porter

Lower Left: TA-3B, BuNo 144856, of VAQ-33.
René J. Francillon

Copyright © 1987 by Aerofax, Incorporated
All rights reserved.
Printed in the United States of America
Library of Congress Catalog Card Number 87-070384
Library of Congress in Publications Data
Francillon, René J.
  Douglas A-3 *Skywarrior*
  (Aerograph 5)
  Bibliography: p.134
   1. Douglas A-3 (Attack Aircraft)
   2. Jet Planes, Military
  ISBN 0-942548-35-3 Softcover
       0-942548-37-X Hardcover

European Trade Distribution by —
Midland Counties Publications
24 The Hollow, Earl Shilton
Leicester, LE9 7NA, England
Telephone: (0455) 47256

THE DOUGLAS A-3
SKYWARRIOR
by René Francillon

CONTENTS

Preface	5
Acknowledgements	5
Chapt. 1: Nuclear Deterrence and the Cold War: The Need for Carrier-Based Strategic Bombers	7
Chapt. 2: Concepts and Design Competition	13
Chapt. 3: Design, Tests, and Evaluation	17
Chapt. 4: The Bomber Variants	25
Chapt. 5: The Versions	31
Chapt. 6: Operational Service	41
Chapt. 7: Unit Histories	55
Chapt. 8: Tales of the *Whales*	85
Chapt. 9: A-3 Technical Description	89
Chapt. 10: Powerplant	119
Appendix 1: Individual Aircraft Histories	121
Appendix 2: Squadron Assignments	134
Bibliography	134
Index	136

Powered by two 7,000 lb. thrust Westinghouse XJ40-WE-3 turbojets, the first XA3D-1, BuNo 125412, made its 30-minute maiden flight at Edwards AFB, California, on October 28, 1952, with George R. Jansen in the left seat and Walter Kent in the right seat. The aircraft is seen here about to touch down at Edwards. Noteworthy are the aircraft's fully opened speedbrakes and extended tail bumper. As various discrepancies forced a postponement of flight trials, the second flight was made more than five weeks later.

By the time he began working on the XVA(H1) Long Range Heavy Attack Aircraft, Ed Heinemann—seen here inspecting the metal mock-up used to test the A3D landing gear—was already one of the most respected aircraft designers in America. His success with the A3D "Skywarrior" and the A4D "Skyhawk" ensures him a place of honor among the world's all-time great aeronautical engineers.

PREFACE

On March 31, 1956, Cdr. Paul F. Stephens, the Commanding Officer of Heavy Attack Squadron One (VAH-1), led a flight of five carrier-based twinjet aircraft from NAS Patuxent River, Maryland, to NAS Jacksonville, Florida. This event marked the initial delivery of Douglas A3D *Skywarriors* to the Fleet. Today, exactly 30 years later, the type's highly successful career is far from over as four Fleet squadrons, two Reserve squadrons, and detachments at NAF Andrews and NAS Point Mugu and at NWC China Lake, still fly this venerable aircraft.

The *Skywarrior's* longevity is particularly noteworthy as it was produced in only limited numbers (the last of 283 aircraft being accepted in January 1961). Moreover, far from living sheltered lives, these aircraft have spent much time at sea and flew combat sorties in Southeast Asia for the duration of the conflict.

Unquestionably, it is owing to the foresight of its chief designer, Edward H. Heinemann, that the *Skywarrior* has outlasted most other types of combat jet aircraft (notably its intended successor, the North American *Vigilante*, which was phased out of service after only 18 years). From the onset of the program Ed Heinemann, one of the world's greatest aircraft designers, insisted that the aircraft be as light as possible, yet strong enough to take the stress of carrier operations and have good growth potential. In 1949 he succeeded in convincing a skeptical Bureau of Aeronautics that the Douglas Model 593 would meet the requirements of Preliminary Type Specification OS-111 for a long-range carrier-based attack airplane capable of delivering a 10,000 lb. nuclear bomb. Indeed, the *Skywarrior* did precisely what the U.S. Navy initially wanted it to do. More importantly, however, it was the ease with which it could be adapted to fulfill other roles that led to operational use long past its originally contemplated service life.

In writing this book I set myself two goals: to communicate my admiration for Mr. Heinemann and to record adequately the history of the "Whale" (the *Skywarrior* remaining to this day the heaviest aircraft to have operated from aircraft carriers). I hope that I succeeded.

As friends in the A-3 community are prone to say "Save the whales, fly A-3's!"

Vallejo, California
March 31, 1986

ACKNOWLEDGEMENTS

Aerofax, Inc. and the author are indebted to numerous individuals and organizations for their generous contributions of photographs and documents used in this **Aerograph**. In particular, we express our sincere appreciation for the assistance provided by the following individuals: Gerald Balzer; Mike Barrett; Michel Benichou; Roger Besecker; Steve Billings; John L. Blazich; Peter M. Bowers; Ken Buchanan; George Cockle; Alain Crosnier; Jean Cuny; Mike Dengler; Fred C. Dickey, Jr.; Jerry Edwards; Robert J. Esposito; Harry S. Gann, Jr.; Jean-Michel Guhl; Michael L. Grove; Warren Hall; Richard P. Hallion, Ph.D.; Dr. Joseph G. Handelman; Chuck Hansen; Fred Harl; Harvey Herzog; E. S. "Mule" Holmberg and Hughes Aircraft Co.; Roger P. Jacobs; George R. Jansen; Clay Jansson; Ed Kane; Karl Kornchuk; A. R. Krieger; William T. Larkins; Mike Laviano; Robert L. Lawson; Peter B. Lewis; Frank MacSorley; E. T. Maloney; Peter J. Mancus; Randy Masters; David W. Menard; Robert C. Mikesh; Cdr. Donald E. Mitchell (CO VAQ-34); Lucien Morareau; Mark Morgan; H. Nagakubo; Stephane Nicolaou; R. T. O'Dell; Douglas D. Olson; D. W. Ostrowski; Alain Pelletier; Ron Picciani; Carl E. Porter; Mervyn W. Prime; Tom Ring; Brian Rogers; Mick Roth; R. C. Seely; Jim Spencer; Jim Sullivan; Arnold Swanberg; William M. Swisher; Norm Taylor: Tommy H. Thomason; Toshiuki Todo; Bruce R. Trombecky; Ens. Kelly Valencia (PAO VAQ-34); John Wells; Gordon S. Williams; Nicholas M. Williams; and Dolly Wilson.

Throughout the preparation of this book we were fortunate to benefit from the assistance of the staff of the following government and private organizations: Grumman Aerospace Corporation; Hughes Aircraft Co.; McDonnell Douglas Corporation; Musée de l'Air; Ames Research Center, National Aeronautics and Space Administration; Audio-Visual Section, National Archives; National Atomic Museum; Department of the Navy (CHINFO; COMNAVAIRPAC; Naval Air Systems Command; Naval Aviation History and Archives; NARF Alameda; USS *Kitty Hawk*; VAK-208; and VAK-308); National Air and Space Museum, Smithsonian Institution; Pratt & Whitney Aircraft, United Technologies Corporation; and Westinghouse Electric Corporation.

Finally, we wish to single out the following persons as their help was substantial and greatly eased our task: Fred W. Burton II (for information on inflight refueling developments); John W. Houch (for unearthing so many old documents and sharing his long experience with the "Whale"); Rick Morgan (for his help in proofing the typescript and saving us from many mistakes); Lawrence S. Smalley (for the loan of many outstanding negatives); Don J. Willis (for sharing his wartime recollection); and our artists: Janine Olivereau, Thierry Van Zandt, and Mike Wagnon.

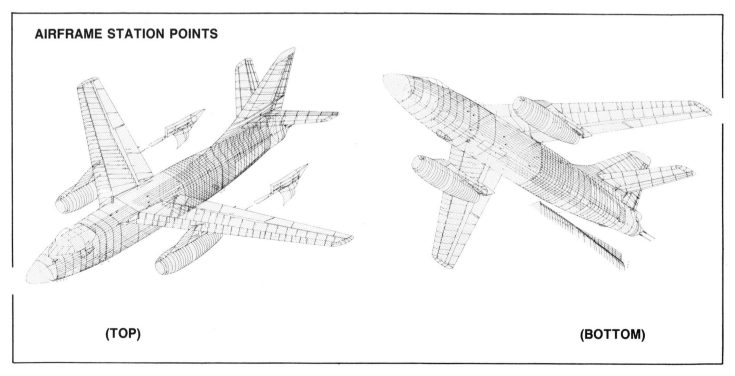

AIRFRAME STATION POINTS

(TOP)　　　　　　　　　　　　　　　　　　　　(BOTTOM)

Sixteen years before the maiden flight of the Douglas XA3D-1, a little-known French aircraft, the Potez 56E, had become the world's first twin-engine aircraft to make arrested landings and takeoffs aboard and from an aircraft carrier. These history-making operations took place aboard the carrier "Béarn" off the cost of southeastern France. In this view the aircraft, flown by Lieutenant de Vaisseau de Lorza, is seen just after catching an arresting cable.

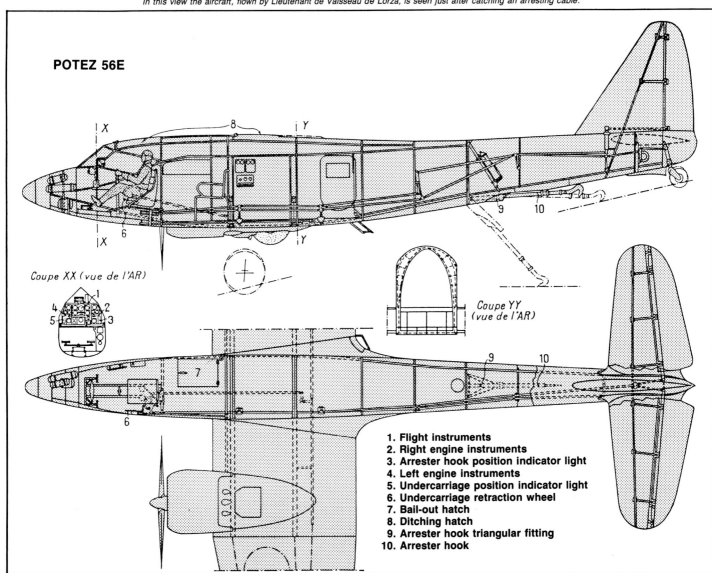

POTEZ 56E

1. Flight instruments
2. Right engine instruments
3. Arrester hook position indicator light
4. Left engine instruments
5. Undercarriage position indicator light
6. Undercarriage retraction wheel
7. Bail-out hatch
8. Ditching hatch
9. Arrester hook triangular fitting
10. Arrester hook

Chapter 1:
Nuclear Deterrence and the Cold War: The Need for Carrier-Based Strategic Bombers

As part of its 1937 proposal to develop a twin-engine, carrier-based fighter, Lockheed recommended that one of its JO-2 light transports be modified and used for carrier evaluation. Fitted with a fixed tricycle undercarriage, this XJO-3 (BuNo 1267) was flown by LCdr. Thurston B. Clark and is seen here aboard the USS "Lexington" (CV-2) on August 29, 1939, during the first test of a U.S. twin-engine aircraft aboard a carrier.

In its design mission—strategic nuclear deterrence—the Douglas A-3 *Skywarrior* was eminently successful. Tests and operational training convincingly proved that the A-3 would have struck its designated targets with reasonable chances of success if the need had ever arisen. The fact that the *Skywarrior* was never required to do so was probably fortuitous. Yet, along with the other elements of the United States strategic forces—SAC bombers and early ICBMs, the A-3 was an effective brake on Soviet expansionism during the peak period of the Cold War. In so doing, and later in its use as a tanker, the *Skywarrior* broadened the role of naval aviation.

At the onset of World War II, carrier aviation was generally regarded as a mere adjunct to the surface fleet, with battleships remaining the primary striking element of the world's navies. The value of carrier aviation in obtaining local air superiority, and in providing a far-hitting tactical force wherever friendly land bases were not available, was nevertheless recognized early in the war; the Royal Navy used carriers in April 1940, during the abortive invasion of Norway, and in November 1940, to strike a swift blow at the Italian Navy in Taranto. Of even further reaching significance in the development of carrier aviation was the attack on Pearl Harbor by aircraft from six fleet carriers of the Imperial Japanese Navy on December 7, 1941. As the war went on, particularly in the vast Pacific, carriers and their aircraft soon supplanted battleships as the cornerstone of naval forces during epic air/sea battles in the Coral Sea and off Midway, as well as in the Solomons and off the Marianas and the Philippines. U.S. carrier aircraft, and to a lesser extent their British counterparts, also established air superiority in the course of scores of amphibious invasions from Guadalcanal to Okinawa, as well as during the landings in North Africa, Italy, and southern France. Yet, through all these actions, carrier-borne aviation retained primarily a tactical role.

The employment of carrier-launched aircraft against industrial targets, a strategic role, was pioneered jointly by the USN and the USAAF on April 18, 1942. On that day, 16 North American B-25B *Mitchell* medium bombers of the 17th Bombardment Group, USAAF, took off from the USS *Hornet* (CV-8) and bombed a variety of targets, including factories and the Yokosuka Navy Yard, prior to reaching China and Siberia. Planned and executed as a morale-boosting effort during the darkest hours of the war, the Doolittle raid achieved its purpose but was of limited military value. Nonetheless, it succeeded in forcing Japan to retain in its home islands a disproportionate number of fighter aircraft, which were critically needed elsewhere by its forces operating over extended lines. More importantly, in carrying a greater load farther than contemporary single-engine carrier-borne aircraft, Doolittle's B-25Bs demonstrated the value of including multi-engine aircraft as part of the complement of carrier air groups.

Credit for the first use of carrier-based aircraft in a genuine strategic role went to a six-nation fleet under Admiral Sir James Somerville, RN, which included the USS *Saratoga* (CV-3) and HMS *Illustrious*. On April 16, 1944, British and U.S. aircraft from these carriers bombed docks and oil storage facilities at Sabang, Sumatra; 31 days later they struck similar targets at Soerabaja, Java. These initial attacks against Japanese oil supplies were repeated by aircraft from British carriers during the next seven months, with the best results obtained during the January 1945 raids on the oil refineries at Pangkalan Brandon (on the 4th), Pladjoe (on the 24th), and Soegi Gerong (on the 29th).

Operating closer to the Japanese home islands, Task Force 38 under Admiral William F. Halsey first bombed Japanese industrial targets on February 16 and 17, 1945. Then, in preparation for Operation *Olympic-Magestic* (the planned invasion of Kyushu), TF38 was again assigned strategic targets on the islands of Honshu and Hokkaido, including communication facilities and factories, as well as airfields and shipping. Strikes were flown for 14 days in July and August 1945, but limited bomb tonnage (maximum bomb-load for contemporary carrier-based aircraft was 2,000 lbs., whereas B-29s operating from the Marianas could each carry up to 12,000 lbs. of bombs) reduced significantly the effectiveness of carrier aviation against these strategic targets. The relative insignificance of these operations was further highlighted on August 6, 1945, when a B-29 dropped a single bomb—with the equivalent destructive power of 13,000 tons of high explosive—over Hiroshima. Three days later another B-29 released a 23-kiloton weapon over Nagasaki. The dawn of the nuclear age quickened the end of World War II, but few people realized then that it also heralded a new era for carrier aviation: nuclear deterrence on the high seas.

With the exception of these brief strategic forays against industrial facilities in Japan, and against Japanese-held oil refineries in the Dutch East Indies, American and British carrier aviation had ended the war as powerful weapons, well-suited to the control of the seas but ill-fitted to the con-

Carrier trials for the North American PBJ-1H (43-4700) took place aboard the USS "Shangri-La" (CV-38) on November 15, 1944. At that time, the PBJ-1H was the heaviest aircraft yet landed aboard a carrier, but the Navy never planned to use this aircraft for regular carrier operations.

The prototype of the Dewoitine D.750 was one of the three types of twin-engine carrier-based aircraft ordered by the French Ministère de la Marine prior to WWII. The D.750 first flew on May 6, 1940, but by then construction of its intended carriers, the "Joffre" and the "Painlevé", had already been suspended.

duct of conventional strategic operations. The latter was the appanage of land-based heavy bombers, with USAAF day raids and RAF night bombings having laid ruin to the Axis war machine in Europe. Likewise, B-29s of the XX and XXI Bomber Commands had dropped not only the two nuclear bombs but also more than 96% of all conventional bombs which fell over mainland Japan.

The U.S. Navy had taken an early interest in nuclear weapons and two of its officers, Capt. William Parsons and Cdr. Frederic L. Ashworth, not only played important parts in the development of the first nuclear bombs but also were Bomb Commanders on the Hiroshima and Nagasaki missions, respectively. Following the end of the war, the Navy did not lose sight of the potential offered by this newly acquired technology. After all, a single bomber carrying a nuclear weapon could cause more damage than a whole fleet of conventional bombers; a factor of major importance when limited deck and hangar space reduced the number of carrier aircraft that could be launched against major strategic targets. The Navy's activities were directed toward participation in the Bikini Atoll tests, development of nuclear reactors to power ships, and the adaptation of the new weapon for use by carrier-based aircraft.

The latter activity met with considerable opposition from the AAF which considered its long range bombers to be the only nuclear delivery system needed by the nation. Moreover, early atomic bombs were both too bulky and heavy[1] to be carried by contemporary carrier aircraft and had to be armed when airborne, thus requiring large aircraft with an internal bomb bay accessible in flight, a feature not incorporated in carrier aircraft of the

[1] *Little Boy*, the weapon used at Hiroshima was 120 inches long and 28 inches in diameter (3.05 m. x 0.71 m.) and weighed 8,900 lbs. (4,037 kg.). The Nagasaki weapon, *Fat Man*, was even heavier (10,800 lb./4,899 kg.) and bulkier (128 inches x 60 inches/3.25 m. x 1.52 m.). Reduction in size and weight were expected but the timescale of such developments was not clear.

time. However, the commissioning of the new CVB (*Midway*-class) carriers, with their longer and stronger deck, enabled Navy planners to consider the possibility of embarking much heavier aircraft, with greater weight lifting capability and an appropriate bomb bay. A still larger carrier—the ill-fated USS *United States* (CVB-58)—was being studied during the immediate postwar period; its anticipated availability in the early fifties would have further raised the limits on aircraft size and operating weight.

While planning for larger and heavier aircraft and for the carriers from which they were to operate, the Navy lobbied resolutely to be authorized to play a significant role in the national policy of nuclear deterrence. The main argument advanced in support of the Navy's contention was that no major target in the USSR was more than 1,500 nautical miles from points where carriers could operate. Hence, Soviet defenses would be seriously complicated by the need to intercept nuclear bombers coming from any point of the compass. Furthermore, the Navy was quick to point out that storing special weapons aboard carriers would avoid the need of obtaining authorization from allied governments to store such weapons on foreign soil.

To overcome the AAF opposition, John L. Sullivan, Acting Secretary of the Navy, wrote to the President on July 24, 1946, that:

> The atomic bombing of Hiroshima and Nagasaki and the first Bikini tests have amply demonstrated that the atomic bomb is the most effective single instrument of mass destruction ever deployed.
>
> The high mobility of the Naval Carrier Task Force combined with its capacity for making successive and continuous strikes in almost any part of the world make this force a most suitable means of waging atomic bomb warfare. Carrier Forces are particularly effective during the early phases of a war when fixed shore installations may be temporarily immobilized by planned surprise attacks in force. Increased range of carrier aircraft, which will shortly be provided by new engines under development, will further increase the areas accessible to attack by carrier based aircraft. Also, the Carrier Task Force can provide a fleet of fighters to escort its bombers throughout their tactical range and thus ensure maximum probability of successful accomplishment of the bombing missions.
>
> In order to enable Carrier Task Forces to deliver atomic bombs, it will be essential to modify carrier aircraft and alter aircraft carriers to provide servicing facilities. This will require advance peacetime preparations. The aircraft carrier is well adapted for modifications to provide the bomb assembly and technical facilities essential to the preparation of the atomic bomb for combat use. Excellent security can be maintained by assembling the bomb within the ship as opposed to assembly in elaborate and obvious installation ashore.
>
> A memorandum of August 15, 1945, issued by you, requires specific Presidential approval for the Navy to receive such information as would be necessary to permit the Navy to prepare for the use of the atomic bomb in warfare.
>
> I strongly urge that you authorize the Navy to make preparations for possible delivery of atomic bombs in an emergency in order that the capabilities of the Carrier Task Forces may be utilized to the maximum advantage for national defense.

After obtaining the tacit approval of the President on November 19, 1946, the Navy was able to proceed with the development of nuclear bomb-carrying aircraft and with the modification of the three carriers in the CVB class to permit the operation of these aircraft. These challenging tasks were not satisfactorily resolved until the entry into service of the A3D-1 *Skywarrior* ten years later.

Only too aware of the relative ineffectiveness of carrier aircraft when operating against hardened submarine pens and industrial targets, in late 1944 Vice Adm. Marc A. Mitscher foresaw the need to develop a new type of carrier bomber which could carry a bombload of up to 12,000 lbs. After becoming Deputy Chief of Naval Operations (Air) in August 1945, Vice Adm. Mitscher continued urging the development of large carrier-based bombers. The payload-range parameter of such an aircraft, which happened fortuitously to coincide with those needed later for carrier-based aircraft adapted to the delivery of the early types of nuclear weapons, clearly demanded a twin-engine configuration.[2] This in itself was not a new development in carrier aviation as the operation of twin-engine aircraft aboard carriers had been contemplated by the Navy as early as 1927-28. Nevertheless, experience with multi-engine carrier-operated aircraft was, to say the least, quite limited at the time of Vice Adm. Mitscher's recommendation and the initial development of naval nuclear bombers.

The world's first twin-engine aircraft planned for

[2] Until 1945, the carrier-based aircraft capable of lifting the heaviest load was the experimental Douglas XTB2D-1. Designed to carry up to 8,400 lbs. (3,810 kg.) of bombs/torpedoes, the XTB2D-1 first flew on May 7, 1945. Only two prototypes were built before cancellation of a small production contract.

First flown on August 29, 1947, the Nord 1500 "Noréclair" was developed to meet a French requirement for carrier-based torpedo bombers. This program and its associated carriers proved financially overly ambitious for postwar France.

The Douglas T2D-1 was the world's first twin-engine aircraft designed to operate from carriers but engine problems prevented its XT2D-1 prototype from being flown aboard the USS "Langley" in April, 1927, as originally planned.

carrier operations was the Douglas XT2D-1. Developed by Jack Northrop and Ed Heinemann from the Naval Aircraft Factory XTN-1, an experimental twin-engine multi-purpose aircraft derived from an initial design for a single-engine carrier aircraft, it made its first flight on January 27, 1927. Less than two months later, this aircraft was shipped to Hampton Roads, Virginia, and then flown to NAS Anacostia for preliminary Board of Inspection and Survey trials. At that time, and under direct instructions from Rear Adm. William A. Moffett, Chief of the Bureau of Aeronautics, plans were made to divert the USS Langley (CV-1) on her way back from Guantanamo to conduct XT2D-1 carrier landing trials, with Lt. Lyon being designated the pilot to conduct the trials. In compliance with this instruction, the USS Langley steamed toward Hampton Roads during the last week of April 1927, while the XT2D-1 was being readied by substituting a conventional undercarriage for the twin pontoons used during initial trials at NAS Anacostia. Troubles with the aircraft's Wright P-2 engine, however, forced the cancellation of this first carrier test with a twin-engine aircraft. Afterward, carrier trials of the XT2D-1 and of production T2D-1s continued to be advocated by a group of naval aviators, but concerns over directional instability in the event of an engine failure during carrier approach and landing proved to be major stumbling blocks. Moreover, while the performance of the Douglas twin-engine aircraft was impressive, its large size limited the number of aircraft which could be embarked. Consequently, from available records it appears that the XT2D-1 and T2D-1s were operated only from shore bases.

Thus it is that the first carrier trials of a twin-engine aircraft were conducted in France on September 24, 1936 when Lieutenant de Vaisseau de l'Orza made several arrested landings and take-offs aboard and from the *Béarn* with a Potez 56E (an experimental development of the twin-engine light transport/trainer monoplane with retractable undercarriage and arresting hook). Sadly, the first pilot to land a twin-engine aircraft on a carrier was killed during that evening while practicing night landings aboard the *Béarn* in a Levasseur PL.7 biplane. Although the proposed production of the Potez 56E to equip the bombing/scouting squadron of the *Béarn* was not realized, its brief trials were found sufficiently promising for the French Ministère de la Marine (Department of the Navy) to order prototypes of two twin-engine torpedo-bombers in 1937: the C.A.O. 600 and the Dewoitine D.750, flown respectively in March and May 1940. Another twin-engine carrier type, the Breguet 810 dive bomber, was ordered in 1939 but was not completed prior to the collapse of France in June 1940.

In the United States interest in twin-engine carrier aircraft was rekindled by Lt. Cdr. Alfred M. Pride[3] when he served successively as Head of Class Desk D (torpedo aircraft) and Head of Class Desk A (fighter aircraft) at the Bureau of Aeronautics. Under his impetus, Design No. 145, an in-house Navy study for a torpedo bomber powered by two Pratt & Whitney XR-1535-92 engines, was evaluated in the spring of 1937 but was found to offer little or no advantages over the Douglas TBD-1 then entering service. During the following year, another twin-engine torpedo bomber design, submitted as an unsolicited proposal by a Massachusetts industrialist, was rejected because it was too heavy and its sponsor lacked the resources required for its development and production.

Meanwhile, a request for twin-engine fighter proposals was sent by the Bureau of Aeronautics on March 18, 1937, but none of the designs submitted by the industry offered sufficient advantages over single-engine fighters to warrant development. However, as part of its proposal, Lockheed Aircraft Corporation recommended that one of its JO-2 light transports be modified with arresting gear and non-retractable tricycle landing gear to allow carrier evaluation of a twin-engine aircraft. This recommendation was endorsed by the Bureau of Aeronautics and resulted in the first U.S. carrier trials with a multi-engine aircraft. On August 29, 1939, the Lockheed XJO-3 flown by LCdr. Thurston B. Clark made 11 landings and take-offs aboard and from the USS Lexington (CV-2).

Still wishing to acquire a twin-engine fighter aircraft, even after the inconclusive 1937 competition, the Navy issued a new RFP in February 1938. This time the results were more encouraging, and Grumman Aircraft Engineering Corporation was awarded a contract for the XF5F-1 prototype. First flown on April 1, 1940, this aircraft soon ran into teething troubles and never went on to be tested aboard a carrier. Nevertheless, in May 1941, Grumman succeeded in obtaining a follow-on contract for the first purpose-built U.S. twin-engine combat aircraft to operate from a carrier, the F7F-1 *Tigercat*.

Shortly after America's entry into the war, the Director of Planning and the Head of Engineering at the Bureau of Aeronautics issued a joint memorandum urging the development of a twin-engine aircraft to perform both the torpedo bombing and scouting missions. Proposals for this new VTSB-class aircraft were received in February 1942 from Douglas, Grumman, and Lockheed-Vega. Although Douglas had studied various twin-engine designs to fulfill the VTSB mission, it was its single-engine version which was ordered in November 1942, as the use of a powerful Pratt & Whitney XR-4360 radial resulted in the XTB2D-1 having better performance than could be achieved with two smaller engines. As a running mate for this Douglas design, the Navy ordered a prototype of the Grumman XTB2F-1 powered by two Pratt & Whitney R-2800 radials but was forced to drop consideration of the Vega V-141 as the Lockheed subsidiary already had too much work for its limited design staff. As it turned out, the Douglas XTB2D-1 did not fly until May 7, 1945, and production of this aircraft was cancelled prior to carrier trials; the Grumman XTB2F-1 made even less progress as it was cancelled in June 1944, prior to completion of a prototype. Another twin-engine carrier bomber, the XTSF-1, was ordered from Grumman in mid-1944, but development of this F7F derivative was dropped within six months. Thus, when Vice Adm. Mitscher began advocating the development of large carrier aircraft, the Navy still had limited experience with carrier compatible multi-engine aircraft.

Up until the mid-war years, Britain had been slow in pursuing the development of twin-engine carrier aircraft. However, the Fleet Air Arm quickly caught up and took the lead, as a modified version of the de Havilland *Mosquito VI* fighter-bomber was successfully tested aboard HMS *Indefatigable* on March 25, 1944. This success led the Royal Navy to order the production of 50 *Sea Mosquito* T.R. Mk. 33s and six *Sea Mosquito* T.F. Mk. 37s (both versions were fully carrier compatible, with folding wings, arresting hooks, and long-stroke undercarriage)[4], as well as the development of a smaller, faster derivative, the de Havilland *Sea Hornet*. Both the *Sea Mosquitoes* and the *Sea Hornets* were readied too late for wartime operations, but the latter saw service with the Fleet Air Arm until 1954.

On November 15, 1944, less than nine months after the British *Mosquito* had become the first twin-engine combat aircraft to be tested aboard a carrier, two American types made their debut aboard the USS *Shangri-La* (CV-38). The first to "trap" was a Grumman F7F-1 *Tigercat* (BuNo 80291), a purpose-designed twin-engine carrier fighter; it was almost immediately followed aboard CV-38 by a specially modified North American PBJ-1H (BuNo 35277), a Navy version of the B-25H, itself a late development of the B-25B *Mitchell* as flown off the USS *Hornet* for the April 1942 raid against Japan. Both types satisfactorily completed their brief carrier evaluations but the F7F went on to be operated from land bases by the Marines, while the Navy found no requirements for a carrier-based version of the PBJ-1H. Significantly, however, the PBJ-1H remained the heaviest aircraft to have made a carrier landing until August 1950 when pilots of Composite Squadron Five (VC-5) completed carrier qualifications with the North American AJ-1 aboard the USS *Coral Sea* (CVB-43). Prior to touching briefly on the development of the AJ *Savage*, the operational forebear of the *Skywarrior*, mention must be made of other twin-engine aircraft carrier operations, both abroad and in the United States.

Postwar, the British pursued the development

---

[3] This far-sighted officer did not lose his interest in multi-engine carrier aircraft; he went on to play a vital role in the development of the A3D when, as a Rear Admiral, he was Chief of the Bureau of Aeronautics between May 1947 and May 1951.

[4] In addition, 29 *Mosquito* B. Mk. IV bombers of the Royal Air Force, which had been fitted to carry *Highball* spinning mines, were equipped with arresting hooks for service with No. 618 Squadron aboard carriers of the British Pacific Fleet. Albeit these aircraft arrived in Australia at the end of 1944, they were never embarked.

Although specially developed as the U.S. Navy's first twin-engine aircraft for carrier operations, the Grumman "Tigercat" was only used for land-based operations by the Marines. This F7F-1 was photographed aboard the USS "Shangri-La" (CV-38) during the type's initial carrier trials on November 15, 1944.

Developed in competition with the Nord 1500, illustrated on page 8, the S.N.C.A.C. N.C. 1070 first flew on May 23, 1947. Of highly unusual configuration, the N.C. 1070 carried its three-man crew and armament in a central nacelle while its engine nacelles were extended aft to support the tail surfaces.

By replacing the N.C. 1070's two 1,600 hp Gnome et Rhône 14R24/25 radial engines with a pair of 5,000 lb. thrust Rolls Royce "Nene" turbojets, in 1948 France obtained the prototype of the world's first twinjet carried-based bomber. The type, however, was never landed aboard or launched from a carrier.

Never intended as an operational type as funds for the construction of the required aircraft carrier were deleted from the meager postwar French budget, the N.C. 1071 was obtained by modifying the piston-powered N.C. 1070 which had been damaged in a wheels up accident during its 30th flight on March 9, 1948.

The N.C. 1071 first flew on October 12, 1948. Although its drag was not negligible as evidenced in this front view, the N.C. 1071 was expected to have a top speed of 494 mph at 21,325 ft. In fact, the aircraft never could exceed 390 mph. Note that the nose wheel was offset to the left.

of the de Havilland *Sea Hornet* and also tested the Short *Sturgeon* twin-engine torpedo-bomber intended for operations aboard the new HMS *Ark Royal* and *Hermes*. However, even the 21,700 lb. (9,850 kg.) gross weight of this latter aircraft was timid in comparison with contemporary French efforts. Not all planning and design activities had been suspended in France during the German occupation; in 1943, the Marine Nationale (French Navy) had sponsored the design of the PA.25 light carrier and of a twin-engine bomber for use from this carrier. Naught came from these surreptitious wartime projects but the French Navy's interest in heavy carrier aircraft did not wane and led to the ordering in 1945 of prototypes of two twin-engine designs, the conventional Nord 1500 and the twin-boom S.N.C.A.C. N.C. 1070. Neither design could proceed past the prototype stage as budgetary limitations resulted in the cancellation of suitable carriers. The second N.C. 1070 airframe, however, was modified into the N.C. 1071 in which the two piston engines were replaced by turbojets, giving birth to the world's first carrier-compatible twinjet bomber. First flown on October 12, 1948, it was anticipated to have a top speed of 494 mph at 21,325 ft (795 kmh at 6,500 m). However, to reach its full potential, the N.C. 1071 would have required a major redesign, an undertaking found unwarranted as available French carriers were unable to accommodate aircraft grossing over 30,000 lb. (13,720 kg). Nevertheless, this unique aircraft occupies a special place in the history of naval aircraft as it flew four years before the Douglas XA3D-1, the type generally accepted as being the world's first twinjet carrier bomber[5].

In the United States, additional experience with handling large aircraft aboard carriers was gained on January 29-30, 1947, when six ski-equipped Douglas R4D-5L transports bound for *Little America* in Antarctica were launched, with the help of JATO bottles, from the USS *Philippine Sea* (CV-47). This feat, however, was unrelated to the flurry of activity which had followed the approval by the Assistant Secretary of the Navy (Air) of a program outlined by Rear Adm. Harold B. Sallada, Chief of the Bureau of Aeronautics, and aimed at obtaining the type of larger carrier aircraft advocated by Vice Adm. Mitscher.

In a letter dated December 11, 1945, and addressed to the Chief of Naval Operations, Rear Adm. Sallada had written:

1. The largest bomb regularly carried by carrier-based aircraft in World War II was the 2,000 lb G.P. bomb, and the maximum striking radius of carrier-based aviation was about 400 nautical miles. Analysis of bombing results in Germany has revealed that lethal damage to many targets required 12,000 lb. bombs. The handicap of inadequate strike radius in carrier-based aircraft in World War II is well-known, particularly in the earlier phases of the war.

2. A series of preliminary studies has been made in BuAer with the objective of determining what extension of range and bomb size in carrier-based aircraft can be attained through technological advances in the foreseeable future. These studies reveal that development of propeller-turbines will permit considerably greater advances in bombload, range, and speed that can be attained in the immediate future with reciprocating engines or combinations of reciprocating engines and turbojets. Insofar as weight and size are concerned, the results of the studies may be divided into three classes of aircraft, as follows:
   A. Those which will operate from present CVB

---
[5] For the record, it is appropriate to note that the first U.S. test of the adaptability of jet aircraft to shipboard operation was made on July 21, 1945 when LCdr. James Davidson made successful landings and takeoffs from the USS *Franklin D. Roosevelt* in a McDonnell FD-1 *Phantom*. Abroad, the first jet to be tested aboard a carrier had been the specially modified second prototype of the de Havilland *Vampire* which, flown by LCdr. Eric M. Brown, went aboard the HMS *Ocean* on December 3, 1945.

class carriers, and remain within maximum capacities in CVB arresting gear, catapults, elevators, flight deck, and hangar deck.

B. Those which will operate from present CVB class carriers in a restricted manner, i.e., incapable of being struck below, but capable of landing in light loading condition and of taking off without assistance in fully loaded condition.

C. Those which will require a new class of carriers, with arresting gear, catapults, elevators, flight deck, and hangar deck of increased capacities, dimensions, and/or strength.

3. Results of studies to date indicate that aircraft employing propeller-turbines can be developed in each of the above categories with approximately the following major characteristics:

	A	B	C
Gross weight, fully loaded (lbs)	30,000	45,000	100,000
Gross weight, landing (lbs)	20,000	30,000	65,000
Bomb size (lbs)	8,000	8,000	8,000
Vmax/alt (mph/ft)	362/SL	500/35,000	500/35,000
Combat radius (naut. miles)	300	1,000	2,000

In each case above, the 12,000 lb. bomb can be carried with sacrifice of fuel and range. Also, an aircraft can be developed in each category above, around the same bombload, employing reciprocating engines and turbojets, but at a very great reduction in speed and range from the above figures.

4. BuAer recommends that a definite program be initiated to extend greatly the limiting ranges and bomb sizes of carrier-based aviation. Such a program would involve not only the development of the necessary aircraft, but also development in the fields of carriers, bombs, ground handling equipment for the aircraft, and bomb handling equipment. If such a program is initiated, BuAer recommends that it include the following:

I. As soon as possible, in order to accelerate complete definition and solution of design and operating problems involved, an aircraft in Category B, employing reciprocating engines and turbojets, with the object of attaining approximately the following characteristics:

Gross weight, fully loaded (lbs)	40,000
Gross weight, landing (lbs)	28,000
Bomb size (lb)	8,000
Vmax/alt (mph/ft) (jets on)	500/35,000
Combat radius (naut. miles)	300

II. When stage of development of propeller-turbines permits, a turbine-propelled aircraft in Category B.

III. Fully coordinated design and development of carrier, aircraft, and all accessories and components in Category C.

IV. Coordinated design and development of escort fighters to parallel that of long-range heavy bombers.

A development program to obtain a VG-class bomber (corresponding to Rear Adm. Sallada's Category B aircraft with reciprocating engines and turbojets) was indeed assigned a high priority. With this authority, Outline Specification OS-106 was drawn up and forwarded to various manufacturers on 25 January 1946 with the request that engineering proposals be submitted by May 1. It is important to note that at that time the VG-class aircraft were intended as conventional bombers; yet, key Navy planners were already aware that the aircraft could probably be adapted for carrying atomic bombs. In the event, following the November, 1946, presidential approval for the Navy to plan for possible delivery of atomic bombs, the winner of the design competition did go into production as a nuclear bomber.

Following evaluation of the VG-class proposals, Contract NO(a)s-8348 for three XAJ-1 prototypes was awarded to North American Aviation, Inc. on June 24, 9146. Design, manufacture, test, and evaluation of the new aircraft would, however, require at least two years and, even when assuming no serious teething troubles, production aircraft

BuNo 17238 was one of six Douglas R4D-5Ls launched on January 29, 1947, from the USS "Philippine Sea" (CV-47) while the carrier was steaming some 660 miles off the Antarctic Continent. Equipped with skis, the R4D-5Ls landed at Little America as part of Operation "Highjump".

The first purpose-designed heavy attack bomber to enter U.S. Navy service was the North American "Savage". NB 6 is an AJ-1 bearing the markings of Composite Squadron Five. VC-5, which had been commissioned in September, 1948, and initially equipped with "Neptunes", made the first "Savage" deployment beginning in February, 1951.

Noteworthy in this view of a North American AJ-2P "Savage" of VJ-62 are the numerous vertical and oblique camera ports beneath the center fuselage and the exhaust for the fuselage-mounted Allison J33-A-10 turbojet. The piston engines were Pratt & Whitney R-2800-48 radials.

Lockheed P2V-3C of Composite Squadron Five (VC-5) takes off from the USS "Franklin D. Roosevelt" (CVB-42) on September 26, 1949. Even with the use of JATO bottles, the heavy "Neptune" needed the full length of the carrier's deck to get airborne, thus necessitating that all other aircraft be kept below deck.

would not be available for service use until 1949. Fearing that in the meantime presidential and congressional support might be waning, Vice Adm. Arthur W. Radford, then DCNO (Air), and Rear Adm. William Parsons, the leading proponent of naval special weapons, were not prepared to wait that long to give the Navy an initial nuclear deterrence capability. Accordingly, efforts were directed toward securing an interim nuclear bomber capable of operating from CVB-class carriers.

In the Lockheed P2V *Neptune* land-based patrol bomber, the Navy possessed an aircraft which, albeit substantially slower than the contemporary Air Force B-29 strategic bomber, had the necessary payload-range performance for a nuclear strike. Moreover, the dimensions of its bomb bay were sufficient to carry a weapon like the *Little Boy* with adequate access for inflight arming. Finally, in spite of its size (span of 100 ft/30.48 m. and length of 77 ft. 10 in./23.72 m.) and gross weight (63,078 lb/28,612 kg.), the P2V-2 version could still be considered for operations from CVBs. By amendment to Contract NO(a)-9272, this potential was translated into hardware when Lockheed Aircraft Corporation was instructed to modify one P2V-2 and 11 P2V-3 airframes into P2V-3C carrier-launched nuclear bombers.

The first carrier demonstration of the *Neptune* was accomplished on April 28, 1948, when two modified P2V-2s piloted by Cdrs. T. D. Davies and J. P. Wheatley took off from the USS *Coral Sea* operating off the Virginia coast. Captain J. T. Hayward later made 16 arrested field landings at Patuxent River and flew carrier approaches to demonstrate the *Neptune's* capability as a shipboard bomber. While generally successful, these tests revealed that the P2V-3C had insufficient yaw and roll control at low speeds to be acceptable for regular carrier operations; moreover, the wing tip to island clearance was found to be marginal for safe landings in heavy sea state. Accordingly, it was decided that the P2V-3Cs would only be launched from carriers, with recovery to be made either at land bases or, in wartime, by ditching alongside ships after nuclear delivery missions. To make the latter safer, hydrovanes were added to improve ditching characteristics.

The fully modified P2V-3Cs, which had a bombing radar in the nose but defensive armament limited to two 20 mm cannon in the tail turret, were delivered to two specially formed units, Composite Squadrons Five and Six, with VC-5 being the first to be commissioned in September 1948. One year later these *Neptunes* were joined in squadron service by the first AJ-1s, but *Savages* did not supplant P2V-3Cs until early 1952. During the intervening years, some of these *Neptunes* were detached to Port Lyautey, French Morocco, for possible assignment to *Midway*-class carriers serving with the Sixth Fleet. In addition to their contribution to the nation's deterrent, the P2V-3Cs made a number of notable proving and training flights including the following:

March 7, 1949: three VC-5 aircraft were launched from the USS *Coral Sea* off the Virginia coast and one of them, flown by Capt. Hayward, dropped a simulated 10,000 lb. (4,536 kg.) bomb near Muroc, California, before returning non-stop to Patuxent River, Maryland.

April 7, 1949: launched from the USS *Midway* (CVB-41) off Norfolk, Virginia, a P2V-3C flew to NAS Moffett Field, California, via Alaska.

February 8-9, 1950: flown by Cdr. T. Robinson and launched from the USS *Franklin D. Roosevelt* (CVB-42), a P2V-3C flew to Panama and then on to San Francisco, covering a distance of 5,060 miles (8,143 km) in 25 hr. 59 min.

April 21, 1950: A P2V-3C piloted by LCdr. R. C. Starkey took off from the USS *Coral Sea* at the record weight of 74,668 lb (33,869 kg).

All P2V-3C carrier launches, be it for these record flights or for operational flights, were made with the assist of JATO bottles and using the full length of the deck. Hence, when operating P2V-3Cs, the deck had to be cleared of all other aircraft, thus reducing drastically the number of aircraft and causing much displeasure among CAGs and skippers. This situation, however, was temporary as, with the entry into service of the AJ-1, the Navy finally possessed a fully carrier-capable nuclear bomber which could be launched and recovered conventionally from the straight deck of contemporary carriers. Only the size of the *Savage*, as would later be the case with the *Skywarrior*, chagrined the CAGs.

First flown on July 3, 1948, the XAJ-1 was unique in being powered by three engines (two piston engines on the wing leading edge and one turbojet in the aft fuselage) and could carry the larger *Fat Man* weapon. Its trials proceeded quite smoothly in spite of the loss of one of the prototypes and its crew. Fourteen months after the maiden flight of the prototype, the first AJ-1 production aircraft were delivered to VC-5. By the end of August, 1950, the VC-5 crews were carrier qualified and the Navy had its first genuine nuclear bomber. The *Savage*, however, had insufficient range (1,500 naut. miles/2,780 km. when carrying a Mk. 15 weapon) and, by the time it was fully operational, its top speed of 449 mph (722 kmh) rendered it obsolescent. In addition, the AJ had numerous shortcomings including weak hydraulic systems, poor engine-out handling characteristics, and a complicated wing folding procedure requiring external installation of hinges and actuators. Nevertheless, the *Savage* (AJ-1, AJ-2, and AJ-2P) pioneered the way for the *Skywarrior*—including later the new air refueling mission—and solidly established the Navy's role in nuclear deterrence.

In an attempt to overcome the anticipated performance deficiencies of the AJ, the Navy ordered from North American two prototypes of the XA2J-1, a derivative of the *Savage* powered by two powerful T40 turboprops driving contra-rotating propellers. The XA2J-1 first flew on January 4, 1952, but serious power plant problems and the appearance ten months later of the first prototype of a jet-powered aircraft designed for the same mission rendered the A2J program stillborn. The tumultuous development of the jet-powered aircraft—the Douglas XA3D-1, sire of the *Skywarrior* family—is recounted in the next chapter.

Built as a seven-place P2V-2 land-based patrol bomber, this "Neptune" (BuNo 122449) was converted as the first five-seat P2V-3C carrier-launched nuclear strike bomber. It is seen here during a manufacturer's test flight after conversion to P2V-3C standard. Its sole defensive armament consisted of twin 20-mm cannon in the tail.

Chapter 2:
Concepts and Design Competition

Artist rendering dated November 24, 1948, illustrating the Douglas Model 593 as first submitted to the Navy as part of the XVA(H1) competition. The high-mounted wing and underslung engines later to be found in the XA3D-1 and production versions of the "Skywarrior" were already apparent. The length of the nose and the size and location of the cockpit were conversely quite different from those later adopted by the Douglas design team.

While the design of the North American AJ-1 was proceeding apace in the manufacturer's facilities in Inglewood, California, BuAer planners in Washington were already contemplating the requirements for its successor (the Category C bomber in Rear Adm. Sallada's letter of December 11, 1945 as quoted in the preceeding chapter). To these planners, it was obvious that performance of the Navy's future strategic bomber would have to exceed markedly that of the XAJ-1. At least in terms of speed and cruising altitude, it would have to come close to the new Army jet bomber prototypes (the Consolidated XB-46, the Boeing XB-47, and the Martin XB-48) ordered as replacements for the wartime Boeing B-29s.

The most promising of these new Army bombers was the Boeing XB-47 which had been ordered in April of 1946, two months before award of the Navy contract for the XAJ-1. Of truly revolutionary design, the XB-47 was powered by six turbojets housed in two double and two single nacelles below wings with 35° of sweep. The XB-47 was expected to reach a top speed of 578 mph at 15,000 ft. (929 kmh at 4,570 m.) and to have a service ceiling of 37,500 ft. (11,430 m.), thus considerably exceeding the calculated performance of the XAJ-1.

The performance shortcomings of the XAJ-1 placed the Navy in double jeopardy. Militarily, the production version of the new North American naval bomber could be anticipated to have difficulty in penetrating a screen of enemy jet interceptors. Politically, the aircraft became the target for lobbyists supporting the claim of the Army Air Forces to have the Strategic Air Command as the sole provider of American nuclear deterrents. In support of their forceful arguments, these lobbyists pointed out that it made little sense to share the limited postwar defense budget between the AAF, which soon would have jet bombers capable of penetrating Soviet defenses, and the Navy, which was planning to use "obsolete" piston-cum-turbojet aircraft as carrier-based nuclear bombers.

The Navy planners' search for a XAJ-1 successor was complicated notably by: (1) the size of early nuclear weapons which required aircraft with large bomb bays and, hence, relatively large overall dimensions; (2) the limited thrust of early postwar turbojets which made it difficult to obtain the necessary performance with a twin-engine design of a size compatible with aircraft carriers; (3) the high fuel consumption rate of these turbojets which tended to dictate still larger and heavier aircraft, with sufficient internal space to house the fuel required to achieve the specified range; and (4) the weight and size limits imposed on the aircraft by the need to operate from carriers, which said limits jeopardized the possibility of coping with the other three factors.

In the lean postwar period, all major aircraft manufacturers were attentively following these BuAer studies and awaiting with interest the impending call for proposals. Perhaps keenest of all was the Douglas Aircraft Company, Inc. Headquartered in Santa Monica, California, Douglas had limited experience with jet aircraft design but had two distinct advantages over most of its competitors: its El Segundo Division was then one of the two leading aircraft suppliers to the U.S. Navy, and the chief engineer of that division, Edward H. Heinemann, was held in high esteem by BuAer personnel. In order to maintain its prominent place among American naval aircraft manufacturers, Douglas closely watched the work of the Navy planners and kept them apprised of its own design activities. Thus, during a trip to Washington, D.C. in November, 1947, Ed Heinemann discussed a proposed naval version of an Army medium bomber then under study at Douglas. Naught came directly from this undertaking but, a few weeks later, Douglas was informally shown Preliminary Type Specification OS-111, dated December 23, 1947, which called for an aircraft with a gross weight of 100,000 lb. (45,360 kg.), fitted with a tail turret but no armor, and capable of carrying a 12,000 lb. (5,443 kg.) *Fat Man* weapon.

While Douglas and the other manufacturers jockeyed for position, the Navy continued to refine its requirements for the new Long Range Carrier Attack Aircraft. Notably, in a confidential letter sent to the Chief of the Bureau of Aeronautics on January 5, 1948, the Deputy Chief of Naval Operations (Air) defined the parameters for this aircraft and stated:

1. There is a requirement for development at high priority of a long range carrier based attack airplane capable of delivering a bomb load of 10,000 pounds (4,536 kg.).

2. There is a parallel requirement of equal priority for optimum provisions for protection, whether by development of escorting aircraft or by outstanding performance of the attack airplane itself.

3. (a) The airplane may be designed for operations from class 6A carrier.

   (b) The operating altitude of the airplane should be 40,000 ft. (12,190 m.) or higher.

   (c) Emphasis must be placed on maximum possible speed. If suitable aircraft escorts appear not to be feasible, a Vmax of at least 525 knots at 40,000 ft. (972 kmh at 12,190 m.) should be sought. Cruising speed shall be the highest practicable.

   (d) Radius of action of 2000 nautical miles (3705 km.) or greater shall be provided, but range considerations must be secondary to speed, except that an absolute minimum radius of 1700 nautical miles (3150 km.) is required.

   (e) Fuel shall be provided for a maximum speed approach to and retirement from the bomb dropping point of a least 100 miles (185 km.).

   (f) Electronic and other equipment shall be provided to permit safe navigation over the route, detection of other aircraft, bombing, identification, countermeasures, and safe-

ty of flight in all conditions of visibility.
(g) Defensive armament other than tail protection is not desired.
(h) Minimum weight and size should be sought.

The DCNO (Air) requested recommendations from BuAer for changes in Preliminary Type Specification OS-111 thought necessary to permit institution of Phase I of the project. Two months later, in his March 10 letter to the Chief of BuAer, the DCNO (Air) acknowledged that design studies were being somewhat limited by restrictions of maximum gross weight and minimum combat radius requirements. Still holding firm on the earlier stated performance parameters, he nevertheless noted the desirability of exploring several variations in order to reach the best possible compromise.

The DCNO specifically recommended that alternative studies be based upon gross take-off weights of 62,000 lbs. (28,125 kg.), 100,000 lbs. (45,360 kg.), 200,000 lbs. (90,720 kg.), and any weights between 62,000 and 200,000 lbs. (28,125 and 90,720 kg.). Moreover, BuAer was to submit variation studies showing:

(a) Difference if plane does not have landing gear.
(b) If plane has landing gear but will be ditched upon completion of mission due to high landing speed.
(c) If landing gear plane is fueled in the air, at carrier, after take-off.
(d) If landing gearless plane is fueled in air, at carrier, after take-off.
(e) If the studies clearly indicate that the speeds, weights and ranges are completely incompatible, then additional studies based upon "drone" and "control" planes should be made. (This variation study to be made separately and not to interfere with basic study and other variations.)

Finally, the DCNO (Air) stated that these studies had to be given high priority since their results would have a direct bearing upon the characteristics of the new 6A carrier (CVA-58, USS *United States*).

While these exchanges between DCNO and BuAer were taking place, Douglas remained active. Having received a copy of a BuAer design study dated February 13, 1948, for a 130,000 lb. (58,965 kg.) aircraft with jettisonable landing gear, Ed Heinemann and his team felt they could do better, particularly since Heinemann doubted that the construction of CVA-58, the first of the proposed super carriers of the class 6A, would proceed to completion. He therefore felt that the new long range attack aircraft, sought by the Navy, would have to be substantially lighter to operate from existing carriers of the *Midway* and even the *Essex* class.

In quest of this goal, when he was in Washington between March 17-27, 1948, Heinemann showed preliminary drawings to BuAer for a 80,000 lb. (36,290 kg.) turboprop-powered aircraft and for a 70,000 lb. (31,750 kg.) turbojet-powered aircraft. The Navy planners' reaction was cool, to say the least, as they had serious doubts on the feasibility of meeting range and payload criteria with such light designs; moreover, the Navy could ill afford to show favoritism when other companies were working on similar proposals. Undaunted, Heinemann returned to El Segundo where, by April 21, he had refined his jet-powered design which now was anticipated to gross 78,450 lb. (35,585 kg.). These company-funded activities were soon to prove beneficial to Douglas, as a request for proposals was issued by the Navy on August 16, 1948.

Prior to the issuance of this RFP, but after it had been drafted, the BuAer staff issued a classified memorandum outlining its current thinking on the problem. Signed by Ivan H. Driggs, Director of the Design Research Division, this July 28, 1948, document provided details on two Navy studies for turboprop-powered heavy attack aircraft. Study DR-65A, based on the use of two Allison T40-A turboprops, had an overload gross weight of 85,000 lb. (38,555 kg.). With the exception of combat radius, which was calculated to be only 1400 nautical miles (2595 km.), DR-65A was anticipated to meet the performance specified by the DCNO (Air) on January 5, 1948. On the other hand, Study

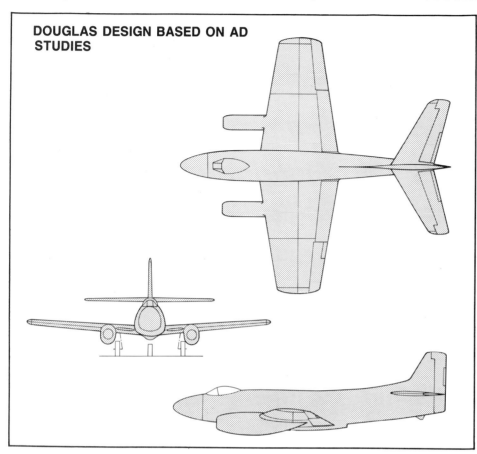

DOUGLAS DESIGN BASED ON AD STUDIES

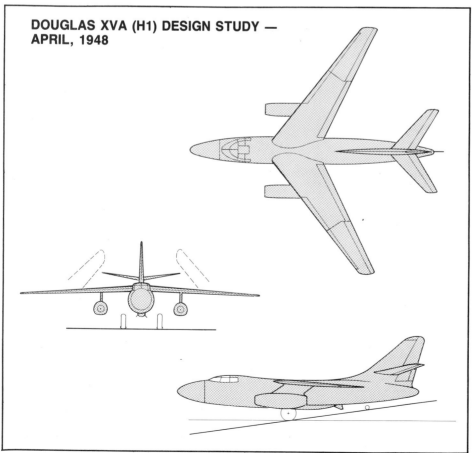

DOUGLAS XVA (H1) DESIGN STUDY — APRIL, 1948

DR-65B, which called for a heavier aircraft (115,000 to 120,000 lb./52,163 to 54,431 kg.) powered by two Allison T44-A turboprops, met all performance requirements. On the basis of these studies, the Design Research Division concluded that the turboprop-powered aircraft would materially exceed the performance of the (DR-64A) turbojet design with respect to combat radius, while retaining practically the same cruising speed and high speed.

As the Request for Proposals for the XVA(H1) Long Range Heavy Attack Aircraft was well advanced, BuAer elected to go ahead in spite of misgivings regarding turbojet-powered designs expressed by its Design Research Division. Accordingly, the RFP calling for a jet aircraft with a gross weight of 100,000 lb. (45,360 kg.) was sent on August 16, 1948, to fourteen manufacturers: Bell, Boeing, Chance Vought, Convair, Curtiss, Douglas, Fairchild, Grumman, Lockheed, Martin, McDonnell, North American, Northrop and Republic.

Upon receipt of the RFP, Heinemann's team at the El Segundo Division set out to win the competition for the Douglas Aircraft Company, Inc. At that time, corporate experience with the production of any type of jet aircraft was virtually nonexistent. True, on May 17, 1946, Douglas had flown the prototype of the first U.S. jet bomber (the AAF's XB-43) but it had not obtained a production contract for this aircraft. Kept out of the wartime design competition which had produced the XB-46, XB-47 and XB-48 Army bombers, the company had undertaken design studies during 1947 for an Army medium bomber and had contemplated the possibility of developing a carrier-based version of this project. Nothing came from either project and, by the summer of 1948, the only jet combat aircraft Douglas had on order were 28 F3D-1 *Skyknight*, carrier-based night fighters. However, both in El Segundo and in the main plant at Santa Monica, the company had highly capable design teams which were pushing the state-of-the-art with work on high-speed research aircraft (the D-558 series for the Navy and the X-3 for the Air Force). Fortunately for Douglas, this limited track record was offset by Edward H. Heinemann's unblemished reputation with the Navy and by his work on various design studies for heavy attack aircraft.

Key personnel on Heinemann's team assigned to the XVA(H1) competition were: Leo J. Devlin, his assistant; I. Eugene Root and Kermit E. Van Every, both aerodynamicists; Dr. William F. Ballhaus, in charge of mathematical studies; R. G. Smith, preliminary designer; and Harry A. Nichols, Project Engineer. It was obvious to this team that if the gross weight were to be 100,000 lb. (45,360 kg.) as specified by the Navy, the aircraft would be limited to operating from only the new CVA-58, then under development. Principally through the sound judgment of Ed Heinemann, it was decided to keep the weight considerably below this limit to enable the aircraft to operate from existing carriers. Accordingly, having calculated that each pound (0.45 kg.) of additional equipment would result in a 6.4 lb. (2.9 kg.) increase in gross weight, Heinemann and his team fought a merciless battle against size and weight.

Starting with their April 1948 design for a 78,450 lb. (35,585 kg.) aircraft, the Douglas El Segundo team began by modifying it to incorporate an enclosed bomb bay, as the need to arm the weapon during flight made the originally planned conformal recess impractical. All thoughts of using a conventional landing gear were dropped in favor of a tricycle undercarriage, even though it meant accepting a weight penalty. Design features were slowly firming up along the lines described in the next chapter. Finally, on November 19, 1948, Douglas submitted technical proposals for three variants—Models 593, 593-1 and 593-2, with cost data being forwarded to the Navy two weeks later.

On December 3, 1948, competitive bids were submitted by Consolidated, Curtiss, Fairchild, Martin, and Republic. The eight other companies which had received the RFP for the XVA(H1) Long Range Heavy Attack Aircraft had declined to bid, either because of other work or because they felt

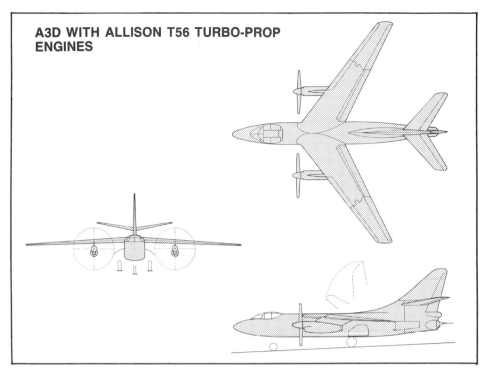

A3D WITH ALLISON T56 TURBO-PROP ENGINES

Labelled "Navy Medium Bomber Study 1A" and dated May 28, 1948, this rendering illustrates one of the earliest concepts evaluated by Ed Heinemann and his design team. At that time the aircraft was planned to be powered by four turbojets in two underslung nacelles. The wing sweep was approximately that adopted subsequently for the XA3D-1.

The Douglas Model 593-2 was one of three configurations proposed to the Navy in November, 1948. It was characterized by the semi-recessed carriage of a single nuclear bomb and the use of fuselage-mounted conventional undercarriage with a tailwheel.

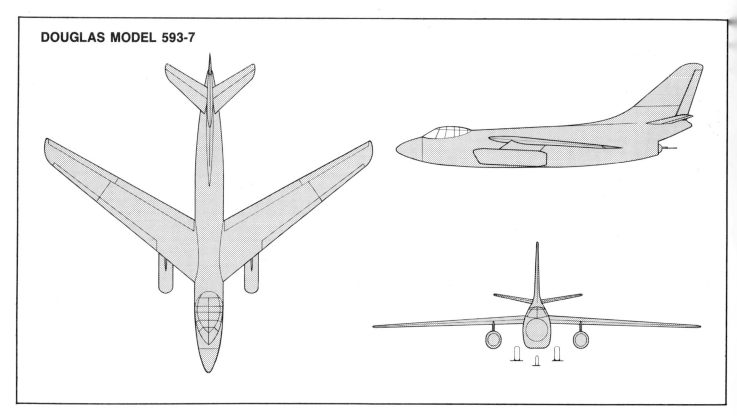

DOUGLAS MODEL 593-7

that the performance requirements could not be met unless the aircraft were to weigh more than what the Navy wanted.

While BuAer reviewed the six proposals, Douglas authorized additional work by Heinemann and his team, resulting in the informal submission of four revised designs: Models 593-3, -4, -5, and -6. After a two-month review, during which the Convair, Fairchild, Martin, and Republic proposals were eliminated, engineering recommendations were transmitted by BuAer on February 2, 1949, and, on March 1, Curtiss and Douglas were requested to submit their costs for Phase I studies. In answer to this request, the two remaining competitors furnished the necessary information, with Douglas basing its March 25 reply on yet another revision of its design, the Model 593-7. On March 31, 1949, Curtiss and Douglas received 30-day Phase I contracts amounting to $808,340 and $810,586, respectively.

At that time, BuAer predicted accurately that a prototype would be flying in October, 1952, regardless of the selection made between the two designs. BuAer's predictions that production aircraft would be available for operation from CVA-58 during the latter part of 1954 were far less accurate. As it turned out, while the two contracts were performing their Phase I studies, news was received that the opponents of large carriers had temporarily succeeded: work on the USS *United States* (CVA-58) had been suspended on April 23, 1949, only five days after the hull had been laid down at the Newport News Shipbuilding and Drydock Company!

The cancellation of the CVA-58 was a serious blow to the Navy's ambition to play a major role in providing a component of the nation's nuclear deterrent, as this carrier was to have provided the primary deck from which heavy attack aircraft were to have operated. For Douglas, on the other hand, it was proof that Heinemann and his team had been right all along in designing as light an aircraft as possible to facilitate operations from the deck of existing carriers. Thus, when on April 29, 194, the CNO instructed BuAer that its ''XVA(H1) Long Range Heavy VA project be re-oriented in order that it may operate from the CVB-41 class carrier,'' Douglas was ready and became the instant favorite as its various Model 593 proposals were all substantially lighter than those from Curtiss.

When news of the CVA-58 cancellation had been received, BuAer telexed Curtiss and Douglas to suspend work on their heavy aircraft projects. This halt only lasted 35 days, however, as BuAer, in compliance with the CNO's instructions, advised the contractors to redirect their studies toward providing aircraft capable of operating from CVB-41 class carriers. The briefly delayed 30-day Phase I studies were completed during the month of May, 1949, with Douglas submitting technical data on its Model 593-8 on the 13th, and cost proposals for the fabrication of two prototypes and a static test airframe on the 31st. Pre-mockup specifications were supplied on June 15, 1949, the revised contract due date, and were accepted by BuAer fifteen days later as full completion on the 30-day Phase I contract.

Prior to the end of the competition, Curtiss and Douglas were asked for one more modification in their designs, as projected weapons necessitated the widening of the bomb bay from 57 inches (1.45 m.) to 66 inches (1.68 m.). Douglas and Curtiss complied immediately, with the former indicating that although its Phase I cost would not be affected by this change, the structural weight of the aircraft would increase.

The final Douglas entry, which provided for an aircraft capable of operating not only from *Midway*-class carriers but also from the smaller, modified *Essex*-class carriers, had little trouble in winning over the substantially heavier Curtiss entry.

During July, 1949, the Secretary of the Navy signed Amendment 1 to Contract NO(a)s-10414, providing $13,017,052 (including a fixed fee of $822,552) for the detailed design and manufacture of two XA3D-1 prototypes and one static test airframe. The saga of the Whale had begun.

The first American jet-propelled bomber design to fly was the Douglas XB-43. Two aircraft were built (44-61508/9) these eventually serving as testbeds at Edwards AFB for a variety of experimental programs. Proposed navalized versions proved abortive. The split canopy configuration remains unique to the XB-43 and several other Douglas designs.

Chapter 3:
Design, Tests, and Evaluation

Substantially lighter than its Curtiss competitor, the Douglas entry won the XVA(H1) competition in June, 1949, and Douglas was awarded a $13,017,052 contract for the detailed design and manufacture of two XA3D-1s. The first of these prototypes is seen here at Edwards AFB where it had been secretly trucked from the El Segundo Division to undergo flight trials away from populated areas and indiscrete eyes.

When its Model 593-8 was confirmed as the winner of the XVA(H1) competition in July, 1949, Douglas Aircraft Company, Inc. was out of its post World War II doldrums and was one of the strongest U.S. aircraft manufacturers. Although the firm had incurred its first ever loss in 1947, it had quickly regained its financial health; in 1948 it realized a profit of $5.8 million on sales of $118.6 million. In Santa Monica, the firm was busy manufacturing DC-6 airliners, while in Long Beach it was getting ready for the production of C-124 military transports. Busiest of all was its Navy-controlled El Segundo Division where production of the AD *Skyraider* was in full swing, the XA2D-1 *Skyshark* and XF4D-1 *Skyray* were taking shape, and the D-558 research aircraft program was pushing the state of the art.

In spite of this flurry of activity, Douglas had the human and industrial resources to undertake successfully a program of the magnitude of that of the A3D. While still competing with Curtiss, Douglas had initiated work on the basic design and mock-up construction in early April, 1949, following receipt of Letter of Intent 02934, dated March 31 and covering the Phase I activities.[1] This work gained impetus during the summer with mock-up inspections resulting in only minor changes affecting primarily the cockpit's arrangement (notably, at the urging of Captain John T. Hayward—the pioneer of *Neptune* and *Savage* operations aboard carriers—engine controls were moved from the left to the center console so that in an emergency they could be reached by the bombardier/navigator in the right-hand seat).

Principal Design Features

As work progressed, with the basic design being completed in February, 1950, and the first engineering releases being made two months later, the XA3D-1 features were firmed up along the general lines already adopted for the Model 593-8 proposal. While the airframe and systems of production aircraft, which closely paralleled those of the prototypes, are described in some detail in Appendix I, it is important to note the reasons behind the selection of the most important features.

The wing location, in shoulder position atop the fuselage, stemmed from the need to provide maximum volume within the fuselage for the large weapon bay. Less easy was the selection of the

---

[1] The cost for the work performed under this Letter of Intent was eventually incorporated into Contract NO(a)s 10414 under the heading R&D Cost to develop Phase I. Altogether, Contract NO(a)s 10414 included the following cost items:

R&D Cost to develop Phase I:	$ 20,419.70
Design studies	4,083.94
Pre-mockup detail specification	334,883.08
Aerodynamic & structural calculations	210,799.10
Wind tunnel testing	187,371.38
Sub-total	757,557.20
Fixed fee 7%	53,029.00
Sub-total	$ 810,586.20
R&D Cost to develop Phase II	
Design data and two prototypes	$12,194,500.00
Fixed fee 6.74527%	822,552.00
Sub-total	$13,017,052.00
Amendments	$ 1,480,760.92
Overrun	$ 1,548,325.77
Total	$16,856,724.89

The first XA3D-1 (BuNo 125412) is shown on approach to Edwards AFB. Like other Navy aircraft of the period, such as the Douglas XF4D-1 "Skyray" and the McDonnell XF3H-1 "Demon", the "Skywarrior" prototype was greatly handicapped by its powerplant, the unreliable and underpowered Westinghouse J40.

XA3D-1, BuNo 125412, shows to advantage its clean wing, fuselage-mounted tail bumper, main undercarriage, and wing slats. Noteworthy are the absence of the tail guns and the extensive use of metal paneling in the canopy design.

Another view of BuNo 125412, in clean configuration during a test flight over the Mojave Desert. The distinctive slanted intake of the J40 nacelles, a distinguishing feature of the "Skywarrior" prototype, is quite noticeable in this view.

wing geometry as speed requirements mandated the use of appreciable sweepback, range requirements dictated to the aerodynamicists the use of a relatively high aspect ratio, and considerations of strength and ease of storage aboard carrier made a low aspect ratio more desirable to stress engineers and operational people. With Ed Heinemann and his Project Engineers—initially Leo P. Devlin and later Harry A. Nichols—serving as referees, a compromise was reached. Sweep at the one-quarter chord line was set at 36°, aspect ratio at 6.75, and wing dihedral at zero degrees, with the latter value being expected to give optimum lateral stability.

Continuing with its previously established philosophy that hydraulically-operated primary controls should have manual reversion for emergency operation, the design team elected to use sufficiently low rudder and elevator boost ratios to allow satisfactory manual control. Inasmuch as a low boost ratio in the aileron system was not possible, dual aileron power systems, completely independent of one another and giving a combined 40:1 boost ratio, were incorporated and provided for a 20:1 ratio in the event of failure of one of the two aileron power systems. In addition, a mechanical aileron control, giving a 2:1 ratio, was provided to enable the aircraft to make a safe field landing after a total failure of the hydraulic boost.

Engine location presented another set of problems but, after considering mounting the engines in nacelles flush with the underwing surface or burying them within the wing root, the design team finally adopted individual nacelles attached to the wing by swept forward pylons. This pod installation not only placed the engines within easy reach of maintenance personnel but also eliminated the need for special servicing equipment as engines could be removed with standard torpedo lifting trucks commonly employed aboard carriers. Moreover, the pods generated minimum drag and facilitated the use of different types of engines, as was eventually the case.

With the selected wing location, the use of a landing gear retracting conventionally into the wings would have required inordinately long main struts lacking the strength required for carrier landings. Hence, the main landing gear was made to retract rearward into the fuselage sides. Greater sensitivity of this close-spaced landing gear to wind direction did not present a problem for carrier operations (launch and recovery being made with the carrier sailing into the wind), but the narrow track concentrated the weight of the aircraft over a very small area. Accordingly, engineers of the Bureau of Ships (BuShips) expressed their concern, and Douglas had to demonstrate with a drop-test rig that the wood deck of *Essex*-class carriers would indeed withstand the landing impact of a heavily laden A3D.

By comparison, the design of the fuselage proved to be fairly straightforward. From nose to tail it was comprised of (1) a large radome containing the Westinghouse AN/ASB-1 navigation and bombing radar, (2) a three-seat cockpit with nose gear and various items of equipment fitted below, (3) a forward fuel cell, (4) a capacious bomb bay accessible in flight from the cockpit, (5) an aft fuel cell above the housing of the main gear, (6) a relatively empty aft fuselage section, and (7) a Westinghouse Aero 21B tail turret with twin 20-mm M3 cannon with 500 rounds per gun. The only unusual feature adopted in its design was the method of emergency escape in flight. Not satisfied with the contemporary ejection seats, concerned with difficulties in designing a satisfactory canopy to provide safe use of ejection seats, and seeking to minimize weight, Ed Heinemann sanctioned the use of a slide-type escape chute as previously employed by Douglas for the F3D *Skyknight* (for details of this chute see Appendix I).

Throughout the *Skywarrior's* life, the lack of ejection seats was the source of much debate and controversy and led to the aircraft's early designation giving place to the unflattering nickname of "All 3 Dead." Indeed, without the benefit of an effective method of emergency egress except when flying straight and level at altitude, the *Skywarrior* had a much higher fatality-to-accident ratio than its contemporaries. As some of his constituents were concerned by the A3D accident rate, Congressman Rousselot queried the Navy and on September 5, 1962, received the following answer from Vice Admiral R. B. Pirie, DCNO:

> The following information is provided in response to your letter of 14 August 1962 concerning the installation of ejection seats in the A3D aircraft.
> The A3D was designed before the ejection seat was a proven means of escape from large, multiple crew jet aircraft and the presently installed escape chute provided the most reliable and simple system for multiple crew evacuation at the time. Unfortunately, although the basic deficiencies of the escape chute have been apparent, the only changes that have been practical and have been incorporated have been minor ones. These involved modification to hatches, hinges, handles and other hardwares in the vicinity of the escape tunnel, and the addition of a static line to provide automatic actuation of the parachute.
> For the past five years the Navy has been trying to find a practical means of improving the crew escape system in the A3D. The results of our protracted and continuing studies have been directed along two primary areas of approach; first, installation of ejection seats and, secondly, development of an escape nose capsule. The incorporation of ejection seats appeared as the most practical and fastest approach and has been thoroughly investigated. An installation, similar to the Air Force RB-66, is possible but represents a major engineering change. The primary factors which have so far precluded the provision of this system have been twofold; first, a significant degradation in aircraft combat performance imposed by the increased weight of the installation and, secondly, the large time-out-of-service required to effect the change which in turn has an adverse effect on operational readiness. The A3D is in critically short supply. Operational commitments and readiness requirements from the present time into the foreseeable future do not indicate any potential alleviation of these conditions.
> It is regretable that combined efforts on the part of the Navy and industry have to date been unsuccessful in providing a new escape system for the A3D which does not adversely affect aircraft tactical performance and critical fleet readiness.
> The Navy is continuing to explore every feasible avenue to improve the means of escape from the A3D which can be incorporated without imposition of unacceptable penalties in aircraft performance and/or combat readiness.

As foretold in Admiral Pirie's letter, the feasibility of installing ejection seats in the *Skywarrior* was explored off and on for the next several years, nearly resulting in a decision to proceed with a retrofit program. However, on March 13, 1969, the Office of the CNO notified the Fleet and the Naval Air Systems Command that:

1. On 14 February 1969, the A-3 escape system was addressed during an informal OPNAV A-3 Program Review. A decision was reached to terminate the A-3 escape sytem retrofit program for the following reasons:
   a. DoD disapproval of funding request.
   b. A low confidence level in manufacturer's production capability.
   c. The long lead time (75 weeks) and a forecast of limited service use.
2. The decision was made reluctantly. A beneficial side effect in expanded technology has been realized that can be applied in follow-on studies. The VSX will have an ejection system and its incorporation has been influenced by the experience gained.

As a result of this decision, the *Skywarrior* will remain the last U.S. fixed-wing combat aircraft not equipped with ejection seats. This anachronism made the news once more when the widow of LCdr. Charles Parker filed a $2.5 million damage suit against Douglas, claiming that the deaths of her husband and his crew were attributable to the lack of ejection seats which prevented them from escaping from their EKA-3B following an operational mishap on launch from the USS *Ranger* (CVA-61) on January 21, 1973. This suit was later settled out of court.

Flutter Studies: The First Bad News

Early in the design of the XA3D-1, aerodynamicists had become concerned with the flutter characteristics of the wing. As a weight savings measure, the aircraft was being designed to load factors considerably lower than those employed in other types of aircraft intended for comparable performance, and it was therefore expected that aeroelastic considerations would be of paramount importance.

Accordingly, two separate investigations were undertaken: a preliminary flutter analysis was performed using the analog computer of the California Institute of Technology (CalTech) in Pasadena, California, and a test program using a flutter model was to supplement and verify the analog analysis.

For the simulation study, Douglas built a mechanical flutter model in which the dynamics of the A3D wing were represented by an equivalent mechanical system which was, in turn, translated into equivalent electrical analog, the force equations being represented electronically. The mechanical system for the wing was a beam in five sections, with coupled bending and torsion degrees of freedom; fuselage and tail masses were included in the inner beam section, rotational inertia and spring terms were used to represent aileron rotation and the engine nacelle was represented by a suspended mass with a rolling inertia and spring.

The electronically expressed force equations for some 1000 cases were analyzed by the CalTech analog computer. The results were positively frightening as it was predicted that flutter would occur at a speed well within the normal operating speed.

Wind tunnel tests, for which a balsa and lead model was used, were performed in the company's tunnel, in the Guggenheim Aeronautical Laboratory of the California Institute of Technology (GALCIT) tunnel, and in the tunnel of the Southern California Cooperative. Unfortunately, these tests confirmed the bad news which had been foretold by the analog analysis.

As predicted by these various studies, the critical mode consisted of wing first-bending coupled with inner-panel torsion, nacelle motion being also considerable. Even with the nacelle pylons made as rigid as practical, flutter became unstable at less than the design limiting speed. Moreover, the wind tunnel tests pointed out another problem: when cruising at Mach 0.8, interference between the nacelle, pylon and wing would cause the Mach number to increase locally to 1.25.

Notwithstanding concerns over this potentially serious defect, on February 10, 1951, the Navy issued a Letter of Intent covering the first twelve production A3D-1's, with the corresponding Contract NO(a)s 51-632 being executed on October 22, 1952. As a result of the Navy's decision to order the aircraft into production, a frantic search for possible palliatives to the flutter problem was initiated. Eventually, relatively minor structural redesign and the strengthening of the engine pylons appeared to yield at least partial solutions. Nevertheless, flight tests of the XJ40-powered prototypes confirmed that the flutter problem had not been fully solved. Fortunately, considerable improvement resulted from replacing the J40s with J57s (the change of engines being undertaken for other reasons as explained below), with the final solution being reached in April, 1954, when the pylons' trailing edge were extended.

The Air Force RBL-X Program

On June 14, 1951, while design of the Navy XA3D-1 was proceeding apace, the Air Force initiated action to select an aircraft to fulfill its RBL-X (Experimental Light Tactical Bomber/Reconnaissance Bomber) requirement. Among the designs considered to replace Douglas RB-26's and Martin RB-57A's in USAF service were newer models of the Boeing B-47, the Martin RB-57, the North American B-45, the Martin B-51, the Vickers *Valiant*, and the Douglas A3D-1.

A proposal for the latter was submitted by Douglas on August 24, 1951, with the manufacturer recommending six major changes to meet Air Force requirements: provision for de-icing, installation of ejection seats, increased load factor, increased bomb load, larger search antenna, and deletion of carrier equipment.

Although not initially favored in some Air Force circles (notably by the Wright Air Development Center which advocated the Boeing B-47 for high altitude missions and the Martin XB-51 for operations at lower altitude), the Douglas submittal was recommended by the USAF Aircraft and Weapons Board on November 29, 1951. This latter recommendation was endorsed on January 12, 1952, and, three days later, the Air Materiel Command notified Douglas that it had won the RBL-X competition.

Designated RB-66 (reconnaissance bomber version) and B-66 (tactical bomber version), the land-based derivatives of the A3D-1 were first ordered on February 4, 1952, when an initial contract for five RB-66A's was awarded. Soon, however, the Air Force began asking for numerous changes, and production RB-66/B-66's bore little but a family resemblance to their naval forebear.

As the Air Force derivative of the *Skywarrior* ended up being virtually a new design, its story is not included in this volume devoted exclusively to the naval aircraft. (A separate title in the publisher's *Minigraph* series covers the full history of the development and operational use of the B-66.) However, before closing this brief mention of the Air Force's interest in the *Skywarrior*, the USAF original intent to order two A3D-1s for test purposes must be noted. In the event, recognizing that the B-66 would differ too much from the Navy version, the Air Materiel Command finally prevailed upon Headquarters not to proceed with the acquisition of the two A3D-1s.

The XA3D-1, the J40-powered Prototype

Early during the preliminary design phase Heinemann and his team had been stymied by the non-availability of engines with sufficient thrust and had therefore considered using four turbojets in twin nacelles. However, more powerful turbojets became available prior to contract award and, in spite of the designer's preference for the new two-spool JT3 turbojet (the future J57 series of military engines) being developed by Pratt & Whitney, the Navy instructed Douglas to use two Westinghouse J40 turbojets. Satisfied with the J34s powering many of its early jets, the Navy had great faith in the ability of Westinghouse Electric Corporation to develop a high-power turbojet and specified the use of J40 engines for most of its aircraft then under development (e.g., Douglas XA3D-1 and XF4D-1, Grumman XF10F-1, and McDonnell XF3H-1). As will be seen later, this decision proved ill-founded and the A3D program—as were the F4D and F3H programs—was saved from failure only after the decision was made to switch power plants.

When the XA3D-1's had been ordered, the program schedule called for the delivery of the first aircraft in July, 1951, with the second following one month later. First flight was then planned for August, 1951. Still intending to meet this fairly demanding schedule, Douglas initiated manufacturing activities in December, 1950, but the company's efforts were soon thwarted by structural and engine problems. Delays resulted from the late delivery of engines, with Douglas receiving the first XJ40-WE-3 only in August, 1951, and were compounded by the need to redesign the wing splice as a result of a static test initiated during the same month. Then, on January 15, 1952, the static test article suffered a major wing failure; fortunately, this did not require a significant redesign

Seen landing following a test hop at Edwards AFB, the prototype XA3D-1 later was stricken from the inventory on May 23, 1956, after being flown to the Naval Technical Training Center in Memphis, Tennessee for use as an instructional airframe.

The Air Force acquired a derivative of the Douglas A-3 in the form of the B-66 "Destroyer". RB-66C, 55-387, is seen following modification at Douglas' Tulsa, Oklahoma facility. Noteworthy are the wingtip antenna fairings and the still functional tail gun.

Like the first prototype, the second XA3D-1 (BuNo 125413) was initially powered by Westinghouse J40 turbojets. Its engine intakes were straight instead of slanted as on BuNo 125412. This aircraft was also the first to be fitted with the twin-gun tail turret initially provided as defensive armament for the "Skywarrior". (McDonnell Douglas via Harry Gann)

BuNo 125413 underwent initial JATO trials at Edwards AFB. First flown on November 2, 1954, the second "Skywarrior" prototype was re-engined with J57s in late 1955 and was used for thrust reverser trials in 1957-58. It can now be seen on display aboard the USS "Intrepid" in New York. (Air Force Flight Test Center via Dr. Richard P. Hallion)

and the fix was achieved with the addition of a mere 20 lb. (9 kg.).

Following completion of the first XA3D-1 (BuNo 125412), which incorporated the latest structural fixes and was powered by two 7,000 lb. (3,175 kg.) thrust XJ40-WE-3 turbojets, preliminary ground trials were initiated in September, 1952, at El Segundo. However, the unreliability of the experimental Westinghouse engines, the erratic behavior of the air turbine auxiliary drive units at normal idle and taxi power, the unpredictability of fall weather in Southern California, and the near impossibility of maintaining the shroud of secrecy surrounding the XA3D-1 if the first flight was attempted in the heavily populated Los Angeles area prompted Douglas to play it safe. The aircraft was therefore trucked to Edwards AFB where it was re-assembled and readied for its maiden flight. On October 21, 1952, high speed taxi runs, during which the aircraft briefly became airborne twice, revealed the need to adjust the aileron boost system. A one-week delay resulted.

At last, 14 months behind the original schedule, the XA3D-1 made its first flight on October 28, 1952. With veteran Douglas test pilot George R. Jansen at the controls and flight test project engineer Walter Kent in the right-hand seat, BuNo 125412 made a smooth 30-minute flight during which it performed to "all expectations."

In spite of this typical public relations pronouncement, all was not well with the aircraft; it was determined that before its next flight the instruments would have to be calibrated, the water-contaminated oil for the auxiliary drive units would have to be purged, and an improved landing gear of required strength would have to be fitted. While this work, which Douglas had hoped to complete in two to three weeks, was being performed in the test hangar at Edwards AFB hydraulic leaks were uncovered. Postponed until December 4, the aircraft's second test hop ended with a torque link scissor breaking as a result of violent nose wheel shimmy at the end of the landing roll. No sooner were the necessary repairs made than yet another 27-day delay occurred. On December 13, 1952, it was found necessary to replace a collapsed fuel cell and to exchange the XJ40-WE-3 engines with J40-WE-6's. Back in the air in January, 1953, the trouble-plagued prototype ran into more troubles, first as a result of the need to switch from Hydrolube to Red oil, and then due to an engine gear box failure. No spare engines were immediately available and, in any case, in March, 1953, the Navy grounded all J40 engines; the J40's were returned to flight status on July 27.

Although the grounding of its powerplant was obviously going to delay both the testing of the XA3D-1 and the investigation of its anticipated high-speed flutter problem, the impact on the overall program was not too drastic as by then it had already been decided that production aircraft would be powered by Pratt & Whitney J57's. As early as January, 1952, the Navy had become concerned with the difficulties experienced by Westinghouse in developing the J40 engine and had begun considering the use of J57's to avoid delaying the urgently needed A3D. Two months later, when it became clear that 7,500 lb. (3,402 kg.) thrust J40-WE-12's, as specified for the production A3D-1's, would not be available prior to the last quarter of 1954, the new Pratt & Whitney two-spool turbojet being developed for the Air Force became an attractive alternative and was given more serious consideration. Even though this would entail the design of revised nacelles and pylons and changes to the aircraft's air conditioning system, the Navy placed an initial order with Pratt & Whitney for J57-P-1 turbojets on April 2, 1952. At the same time, the Navy began processing a change order to Contract NO(a)s 51-632 to have the A3D-1's J40-WE-12 engines replaced by J57-P-1's or P-6's. Moreover, the Navy also specified that all production Skywarriors were to be delivered with J57's. For Ed Heinemann and his team, this decision represented a final vindication of their long advocated recommendation for the use of Pratt & Whitney J57's to power their heavy attack aircraft. Moreover, the new engines would facilitate their task as the additional thrust and lower fuel consumption of the J57 ensured that climb and range performance guarantees could be more easily met.

While J57's were being fitted to the first production A3D-1, BuNo 125412 did not fly for four months. Its J40 engines were grounded, and it was only in August that the first prototype was returned to flight status to undergo Navy preliminary evaluation beginning on the 24th. Also J40-powered, the second XA3D-1 (BuNo 125413) first flew on October 2, 1953, while it was ferried from El Segundo to Edwards. Externally, this aircraft differed from the first prototype in having straight air intakes; those of BuNo 125412 were canted back, as it had been feared that straight inlets (i.e., perpendicular to the line of flight) would not be efficient at high angles of attack.

Significantly, the second XA3D-1 was preceded in the air by the first J57-powered A3D-1 (BuNo 130352) which had first been flown from El Segundo to Edwards AFB on September 16, 1953. Thereafter, the J40-powered prototypes lost much of their importance as, from then on, flight trials were primarily performed with J57-powered aircraft. In fact, the re-engining of the XA3D-1's with J57's had been considered as early as June, 1953, but was only confirmed in principle on January 12, 1954, when a Change Notice was issued. The need to proceed with the flight trials without undue delay, however, resulted in the temporary postponement of parts of this re-engining program. BuNo 125412 fist flew with J57-P-1's on November 2, 1954, and BuNo 125413 received its J57's in late 1955.

Before detailing later events in the test and evaluation program of the J57-powered A3D-1's and subsequent models, it is appropriate to relate key events in the trials of the two XA3D-1's from the fall of 1953 until they were stricken, as well as some of the experiments which were contemplated during that period.

Although primarily used in expanding the type's flight envelope and exploring its low flutter limits, BuNo 125412 was also evaluated by the Air Force in October, 1953, as part of the B-66 development program. It received new nacelle struts in late 1953/early 1954 in an attempt to correct is high-speed flutter problem which, it had been confirmed, originated from the engine pylon-nacelle configuration. At the same time, its vertical fin was reworked prior to its use in January, 1954, for yaw damper ground development work. The first prototype remained at Edwards AFB until the summer of 1954 when it was returned to El Segundo to be fitted with J57-P-1's. Airborne again on November

The YA3D-1, BuNo 130352, the first J57-powered "Skywarrior", and its contemporary, the first production A4D-1 "Skyhawk", BuNo 137813, are seen during a break in test flying at Edwards AFB. Noteworthy is the fact that the YA3D-1's right inboard pylon and nacelle are painted white and covered with wool tufts for airflow studies. First flown September 16, 1953, when it was ferried from El Segundo to Edwards AFB, BuNo 130352 later served at the NMC Point Mugu, prior to being struck off at NARF Alameda during June, 1965.

2 of that year, the first XA3D-1 was transferred to the Naval Air Test Center at Patuxent River on November 5 where it was to serve for accelerated J57 engine service tests. More than one year after BuNo 125412 arrived at PaxRiver, leaks were uncovered in its wing fuel tanks. As a result of this discovery, all A3D's were subject to flight restrictions which required that integral wing tanks not be fueled and temporarily limited available fuel to that contained in fuselage tanks. Fortunately, Douglas soon was able to devise corrective measures and this restriction was quickly lifted for all *Skywarriors*.

Soon after entering flight trials, BuNo 125413 was used for bomb dropping tests and armament demonstrations. These tests, which were repeatedly delayed by recurring problems with the J40 engines, soon revealed marked bomb bay buffeting. Fortuitously, the addition of a lower flat plate spoiler (set for 40° deflection and located forward of the bomb bay) showed excellent promise in reducing this buffet. Further tests, however, were delayed by the need to rework the aircraft's wing fold mechanism. This work was performed during the last week of February, 1954, but did not prevent damage to BuNo 125413's wing during a folding test on March 2, with a resultant minor delay in flight activities. Following the addition of longitudinal slots in the bomb bay spoiler, buffet tests were satisfactorily completed in mid-April 1954, when bomb bay operation was satisfactorialy demonstrated at Mach .9 at 26,000 ft. (7,925 m.). During these tests, an inert T62 store was dropped at 285 knots at 12,000 ft. (528 kmh at 3,660 m.) with separation being confirmed as very clean. From June 26 until October 16, 1954, BuNo 125413 was at the factory where its AN/ASB-1A bombing radar was installed and its bomb ejector platform reworked. Subsequently, the second prototype was returned to flight operation, and in November, 1954, it conducted further bomb dropping tests over the Aberdeen Bombing Range at Edwards AFB. Initial results, including operation of the ASB-1A radar, were judged satisfactory.

As part of its normal development work in direct support of the program for the A3D and that of its closely related cousin, the B-66, and with an eye to achieving significant performance improvements in later versions of its forthcoming DC-8 jetliner, Douglas conducted a theoretical study program of boundary layer control (BLC). To validate the results of this study, the manufacturer attempted to gain governmental support for full-scale wind tunnel tests and, hopefully, a limited flight test program using a modified XA3D-1. With the Air Force showing interest in incorporating BLC in the B-66 and the National Advisory Committee for Aeronautics (NACA) anxious to study the problem, Douglas obtained authorization from the Navy to modify BuNo 125413 for use in this program. During the second quarter of 1955, the Navy and NACA Ames Laboratory conferred to define an evaluation program for the Douglas BLC system. In early June, agreement was reached whereby BuNo 125413 was to be used for full-scale wind tunnel tests in the Ames 40 x 80 ft. tunnel between July and December 1955. It was further agreed that it this initial phase confirmed promising computed results flight tests would then be programmed and that, at a later stage, it would be determined whether or not to proceed with a supersonic blowing type of BLC investigation. However, lack of funding killed the flight test phase of the BLC research program.

One more unusual development program was planned beginning in February, 1956, when it was proposed that BuNo 125413, by then re-engined with J57's, be used for testing thrust reversers. Aerojet and Marquardt submitted proposals for thrust reversers in mid-1956 and in January, 1957, Marquardt was given a letter of intent for design and fabrication of thrust reversers to be fitted to two J57-P-10's (the program was later extended to include two more J57-P-10 turbojets). The testing was conducted at NAS Point Mugu, California, by Marquardt and Douglas and was terminated in 1958.

The thrust reverser test program was the swansong of the *Skywarrior* prototypes. By then, with the type about to become operational with heavy attack squadrons, the XA3D-1's had reached the end of their useful life and were soon to be stricken. Accepted by the Navy on November 5, 1954, and immediately transferred to the Naval

After the decision to re-engine the "Skywarrior" with the new two-spool Pratt & Whitney turbojet then being developed for the USAF, the Navy acquired a small batch of J57-P-1s for use in the A3D development program. BuNo 130352 is seen here fitted with a pair of these reliable turbojets.

BuNo 130352, the lead ship of the J57-powered version, is shown after installation of its tail guns. A coat of temporary water-soluble white paint had by then been applied to facilitate photo recording of some tests activities. Note the outlined NAVY markings on the aft fuselage above the national insignia. (McDonnell Douglas via Harry Gann)

The second A3D-1, BuNo 130353, on display at Edwards AFB during Armed Forces Day 1956. Note the early style of bomb-bay spoiler and the aircraft bright paint scheme with dayglo red panels on the tail and navy blue leading and trailing edges. As usual, the steel aft portion of the nacelles are left unpainted. (L. S. Smalley via William T. Larkins)

Air Test Center at Patuxent River, BuNo 125412 remained there until retired on February 17, 1956. Having flown a total of 179 hours since October, 1952, the first prototype of the *Skywarrior* was stricken from the inventory on May 23, 1956 after being flown to the Naval Air Technical Training Center in Memphis, Tennessee, for use as a training airframe. The second prototype, BuNo 125413, had an even less significant career as it was never officially accepted. However, it did survive and can be seen on the deck of the USS *Intrepid* in New York. Fate had not been kind to the prototypes of one of the most epochal naval aircraft ever built. *Sic transit gloria mundi* (so fades away the world's glory).

The Test Program for the J57-powered Aircraft

To expedite the *Skywarrior* development program, the Navy first obtained a small batch of J57-P-1's and J57-P-1A's taken from an Air Force contract. Long-term requirements for powering production A3D-1's were to be met by J57-P-6's, J57-P-6A's, and J57-P-6B's ordered under Navy contracts (a total of 104 of these engines were eventually acquired), whereas the A3D-2, A3D-2P, A3D-2Q and A3D-2T versions were to be powered by J57-P-10's (a total of 592 of which were procured under Navy contracts). Ratings and principal differences between the various J57 versions powering *Skywarriors* are provided in the accompanying table.

Douglas received its first J57-P-1's on June 1, 1953, and immediately proceeded with the installation of the new powerplant. Heavier and of greater diameter than the J40's, the J57's had to be housed in nacelles of revised geometry with the tip of their intake located farther forward and slightly higher than those of the J40 nacelles. Moreover, the J57's imposed a redesign of the pylons to accommodate the larger nacelles, modified plumbing, and new attachment points.

PRATT & WHITNEY J57

ENGINE MODEL	AIR FORCE	TURBOJET	TURBOJET	TURBOJET	TURBOJET	TURBOJET	TURBOJET
	NAVY	J57-P-1	J57-P-1A	J57-P-6	J57-P-6A	J57-P-6B	J57-P-10
SPECIFICATION	NUMBER	A1632	A1632 App A	N1671		N1671 App A	N1700
	INITIAL DATE	3-10-51		7-3-53		7-3-53	9-2-54
RATINGS (THRUST AT S.L.) POUNDS	TAKE-OFF	10000 Max	10000 Max	10000 Max	10000 Max	10000 Max	10500 Max
	MILITARY	9500 Mil	9500 Mil	9500 Mil	9500 Mil	9500 Mil	10500 Mil
		8250 Normal	8250 Normal	8250 Normal 7400 90% NR 6200 75% Normal	8250 Normal	8250 Normal	9000 Normal
	NORMAL CRUISE	7400 90% NR 6200 75% NR	7400 90% NR 6200 75%NR		7400 90% NR 6200 75% NR	7400 90% NR 6200 75% NR	8100 90% NR 6750 75% NR
FUEL	GRADE	MIL-F-5624	MIL-F-5624	MIL-F-5624	MIL-F-5624	MIL-F-5624	MIL-F-5624
CURVES	NUMBER	T-1164	T-1164	T-1226	T-1226	T-1226	T-1222
WEIGHT, DRY	POUNDS	4160	4160	4125		4125	4200
JET NOZZLE	TYPE	Fixed	Fixed	Fixed	Fixed	Fixed	Fixed
FUEL CONTROL	MODEL	HSD JFC3 Elec.	HSD JFC12 Hydro-mechanical	HSD JFC12 Hydro-mechanical	HSD JFC3 Elec.	HSD JFC12-5 Hydro-mechanical	HSD JFC12-5
DRAWING NUMBERS		181601	181601	229701		229701	258201
DIMENSIONS (Room Temp.) INCHES	DIAMETER MAX. FROM C.L. LENGTH	41.0 27.3 159.3	41.0 27.3 159.3	41.0 27.3 159.3		41.0 27.3 159.3	40.50 27.25 157.52
TYPE CERTIFICATE	NUMBER	MILITARY	MILITARY	MILITARY	MILITARY	MILITARY	MILITARY
QUANTITY				38		66	591
AIRPLANE		A3D-1*	A3D-1	A3D-1	A3D-1	A3D-1	A3D-2 A3D-2P A3D-2Q A3D-2T
NOTES		*First flight 9-16-53. Production for YJ57-P-3. Steel compressor similar to JT3D.	Similar to J57-P-1 except fuel control.	Navy version of J57-P-1. Steel compressor. Airframe manufacturer supplied oil tank.	J57-P-1 engines modified by Douglas. Navy assigned designation.	Similar to J57-P-6 except incorporates 2 strut high-pressure bleed system.	Steel compressor. External features of J57-P-6B. Higher ratings. Similar to J57-P-29W except no water and no oil tank.

Source: Pratt & Whitney Aircraft, Engine Listings, March 15, 1972.

As the construction of the first A3D-1 (BuNo 130352) was not as advanced as that of the second XA3D-1 (BuNo 125413), the former became the first *Skywarrior* to be fitted with the new nacelles and pylons. Temporarily designated YA3D-1 to reflect its status as the lead ship of the J57-powered version, BuNo 130352 made its maiden flight from El Segundo to Edwards AFB on September 16, 1953 (according to the original schedule, the first A3D-1 should have flown in November, 1952, on the power of two J40-WE-12's). As previously indicated, the J40-powered second XA3D-1 actually first flew 16 days after the first J57-powered A3D-1.

In late 1952 and early 1953, several months prior to the completion of the first A3D-1, the Navy and Douglas agreed to incorporate several changes in the production aircraft. The most important of these changes were the redesign of the canopy framing and glazed area to improve visibility, the use of an aileron dual power system in place of the single system of the XA3D-1's, the relocation of catapult attachment points and fittings to provide for level launch, and the enlargement of the bomb bay to accommodate more conventional ordnance or larger special stores. The first two changes were incorporated in all A3D-1's whereas the modifications required for the revised launch procedure and the larger bomb bay were introduced on BuNo 135407, the 13th production aircraft.

From the start of the A3D-1 trials, it was evident that the J57's were more reliable than the J40's and that their installation in enlarged and relocated nacelles did much to solve the *Skywarrior's* flutter problem. Nevertheless, some teething troubles were encountered with the engines (e.g., several engines were initially rejected and the J57's were grounded for 28 days in February/March 1954) and their installation proved at first deficient (e.g., oil and fuel tank leaks had to be corrected and the nacelles and pylons had to be reworked several times to eliminate high-speed buffet with the aircraft finally demonstrating in April, 1954, that it could be flown at Mach .92 with no buffet). Other problems uncovered during the early phase of the flight test program were insufficient lateral control at high speeds (first encountered during the XA3D-1 trials), high-speed aileron buzz, and inadequate low-speed rate of roll. Various fixes were evaluated, including the trial installation of wing fences and aileron snubbers, but it was the addition of wing spoilers which finally solved these problems.

Significant dates and events in the manufacturer and customer evaluation of the A3D-1 are as follows:

May 1954: The Navy conducted its second preliminary evaluation of the *Skywarrior* at Edwards AFB to check fixes for dual-powered ailerons, high-speed nacelle buffet, rate of roll improvements with wing fences, and bomb bay buffet.

July 1954: As indicated earlier, the first A3D-1's were initially flown with J57-P-1 and J57-P-1A engines (the latter featuring a JFC-12 hydromechanical fuel control system in place of the JFC-3 electronic system of the P-1's). These early engines, however, did not fully meet the requirements of the aircraft, and Douglas modified a number of the P-1's into J57-P-6A's which closely resembled the J57-P-6B's being developed by Pratt & Whitney for the production aircraft. J57-P-6A's were first fitted to BuNo 130353 in July, 1954, when this aircraft was modified by Douglas.

September 1954: The third Navy evaluation took place at Edwards AFB to verify various improvements introduced over the preceding five months. Between 27 and 30 September, BuNo 130353, the second A3D-1, flew 14 times, revealing that the fuel consumption rate during the cruise-climb part of the flight was 25% higher than predicted. Preliminary evidence of high fuel flow indicated that early accurate evaluation of the range characteristics was required.

October 1954: The first A3D-1 was returned to the plant to be fitted with its Hytrol anti-skid system and for some structural rework.

Douglas test pilot Bill Davis performed local carrier suitability trials at Edwards AFB where the A3D-1 made high sink rate simulated deck landings.

January 1955: Flight tests revealed a loss of elevator effectiveness at high Mach (above M .97) due to twisting of the stabilizer.

May 1955: On the 3rd, BuNo 130361 set an unofficial transcontinental record by flying from Los Angeles to New York in 4 hours 4½ minutes.

On the 14th, Board of Investigation and Survey (BIS) trials began with the arrival of a first A3D-1 at NATC Patuxent River. A second BIS aircraft, primarily intended for stability and control tests, arrived at Patuxent River on May 20, with a third following on June 23 for use in armament tests.

June 1955: While being flown back to El Segundo for wing spoiler installation, the second BIS aircraft (BuNo 130357) ran off the runway at Kirtland AFB on June 30. The accident occurred after both ADU's failed on take-off, resulting in the loss of hydraulic pressure and electrical system power. Both main wheels had to be replaced and superficial damage to the wing tip repaired prior to restoring the aircraft to flying status.

July 1955: BuNo 135409, the aircraft used by Douglas for wing spoiler testing, demonstrated complete disappearance of undesirable vibrations after being fitted with a revised stabilizer.

August 1955: Tests leading to carrier suitability trials were begun with Field Carrier Landing Practice (FCLP) being initiated with BuNo 130354. Level launch tests were also started with BuNo 135408.

September 1955: Pilots complained of discomfort and fatigue due to unsatisfactory seat design. Corrective measures included supporting the parachute back pack to remove its weight from the pilot's shoulders and modification of the seat to lessen side slippage. Later on, crew comfort and safety were improved when a thin pack Pioneer personal parachute was adopted; this new parachute eliminated premature opening during high-speed bailout. Further improvement in crew comfort came from the use of a K-1 helmet which did much to overcome a troublesome cockpit acoustical contamination problem.

October 1955: On the 3rd, NATC submitted a report indicating that the A3D-1 roll-out requirement for emergency landing at 65,000 lbs. (29,485 kg.) was 7,500 ft. (2,285 m.). As this was judged excessive, Douglas was requested to develop a drag chute system. The initial proposal for the interim deceleration chute was unsatisfactory as it would have necessitated the removal of the tailhook, thus precluding carrier operations and the use of emergency shore arrestment gear when the chute was fitted. Later in the month, Douglas submitted a satisfactory proposal providing for the installation of a drag chute within the aircraft on the tail turret access door with the pennant retained firmly under the outer hull.

In the same report, NATC noted an immediate need for a fuel jettisoning system, improvements in the fuel transfer system, and reduction of elevator force. In January, 1956, after testing several methods for jettisoning and transferring fuel, Douglas successfully tested systems meeting all requirements. In April, 1956, Douglas

BuNo 130355, the fourth production aircraft, in the markings of the Naval Air Test Center at NAS Patuxent River. This aircraft was later modified to become the sole "Skywarrior" to be powered by the more powerful J75 engine and was finally struck off at Litchfield Park, Arizona, in May, 1965.

Photographed at Dayton, Ohio, in September, 1954, the fifth A3D-1 (BuNo 130356) subsequently became the sole YA3D-1Q. As an A3D-1Q, it served with VQ-2 until lost in an accident at Port Lyautey (now Kenitra AB), Morocco, in October, 1958.

demonstrated an increased elevator boost system (with a 7-to-1 boost ratio instead of the original 4.5-to-1 ratio), thus solving that problem.

On the 25th, the Fleet Indoctrination Program (FIP), which had been scheduled to commence during the month using six A3D-1's and two A3D-1Q's, had to be delayed pending correction of the following unacceptable items: (1) high elevator friction and break-out forces; (2) unsatisfactory fuel jettisoning and transfer; (3) cockpit contamination; and (4) compressor stalls. The minimum necessary changes were incorporated beginning in November.

On the 29th, BuNo 130359 crashed near Edwards AFB; Bill Davis and his Douglas test crew were killed. As a radome failure was considered to be a contributing factor, a requirement for preflight inspection of the radome and a 400-knot (740 kmh) IAS restriction were temporarily established. Changes in the radome fastening procedure soon permitted raising the limitation to 480 knots (889 kmh); thereafter this restriction was lifted altogether. Other suspected contributory causes were wing fuel tank leaks and a possible failure of the forward fuselage section on the right-hand side.

December 1955: As the AN/ARN-6 radio compass initially installed proved to be unsatisfactory, the possibility of incorporating TACAN was investigated by the Navy. Eventually, most *Skywarriors* received the AN/ARN-21 TACAN.

January 1956: The need to avoid visual detection due to contrails led to a modification of the bombing periscope to detect contrails, thus enabling pilots to seek flight levels at which the aircraft could operate without unwanted condensation.

On the 31st, Cdr. Frossard ferried BuNo 135421 from Los Angeles to NAS Jacksonville, Florida, in 3 hours 29 minutes.

February 1956: The SR-38 Part I demonstration (manufacturer's trials in accordance with Navy procedures) was completed.

February-March 1956: When A3D-1 flight trials began, the Fleet Indoctrination Program (FIP) had been tentatively scheduled to commence during the early fall of 1955. Aircraft delivery slippage and a number of discrepancies requiring corrective measures delayed the FIP start until November, 1955, and then to February, 1956. Using six A3D-1's and two A3D-1Q's, the three-mongh FIP program finally started on February 12. One of the FIP aircraft (BuNo 135242) crashed at Edwards AFB on February 14 due to the inadvertent extension of the undercarriage during a high-speed run. Parts of the main gear broke off and struck the stabilizer, forcing the crew to bail out. In spite of this accident, the FIP progressed satisfactorily and was concluded two weeks ahead of schedule. Only a few changes were recommended, thus clearing the A3D-1 for service use.

February-August 1956: The Navy assigned BuNo 130352 to *Project RedWing* for participation in nuclear and thermonuclear weapon tests at Eniwetok. This aircraft was damaged on February 14 when its pilot failed to notice the landing gear handle in the up position when starting engines, and a replacement aircraft (BuNo 135434), quickly readied by Douglas, arrived at Eniwetok in April, 1956. Following numerous brake difficulties (the brakes had to be used repeatedly due to the short length of the Eniwetok runway), BuNo 135424 was ferried back to Hickam AFB, Hawaii, for brake overhaul and modification prior to being returned to the Pacific Proving Grounds. After completing its assigned task, the aircraft was flown back to NASWF Albuquerque, New Mexico, covering the last leg (2,801 nautical miles/5,187 km.) in 5 hours 50 minutes; the aircraft still had a 45-minute fuel reserve upon landing. During *Project RedWing* the aircraft also demonstrated its endurance by flying a six hour and eight minute sortie at which time it had a 30-minute fuel reserve.

March 1956: On the 31st, five aircraft were released from the FIP program and delivered to VAH-1 at NAS Jacksonville, Florida, to initiate the *Skywarrior* operational life. Two days later, the two FIP A3D-1Q's were sent to O&R Norfolk, Virginia, to be brought up to fleet delivery configuration. The last FIP A3D-1 was retained at Patuxent River for additional service tests.

March-June 1956: By the middle of 1956, with the BIS trials and FIP program underway and the FCLP tests completed, the A3D-1 was ready for carrier suitability trials. Two aircraft (BuNo 135408 as the primary aircraft and BuNo 135411 as the backup) were instrumented, and the Navy's newest carrier, the USS *Forrestal* (CVA-59) which was then completing its shakedown cruise, was set to receive the *Skywarrior*.

Carrier trials began on April 3 and the first results were successful. The aircraft was at home in this medium with a minimum catapult end speed of 102 knots (189 kmh) reached at 46,000 lbs. (20,865 kg.). Touch-and-go landings and arrested landings were taken easily in stride. Minor deck handling procedures and discrepancies were disclosed such as: (1) failure of crews to disconnect nose wheel scissors prior to towing; (2) failure to re-position center-of-gravity valve causing excessive aft c.g.; (3) need to provide friction coating for the hangar deck; (4) requirement for shorter and stronger tow bar; (5) need to install D.C. power leads from Huff tow tractor to provide brake power when engines were not running; etc. The A3D-1 carrier suitability trials aboard the USS *Forrestal* were satisfactorily concluded on April 23. These had included launches at 70,000 lbs. (31,751 kg.) with full forward and aft c.g., minimum end speed under these conditions being 130 knots (241 kmh), and recoveries with the aid of a temporary mirror landing system.

To clear the A3D-1 for operations aboard 27C Class carriers (modernized *Essex* carriers), BuNo 135408 was then ferried to the West Coast where carrier suitability trials were undertaken aboard the USS *Bon Homme Richard* (CVA-31) from June 18 to 21.

August 1956: The Board of Inspection and Survey reported the completion of the A3D-1 trials.

January 1957: On the 29th, NASWF Albuquerque, New Mexico, submitted a final report on the A3D-1 Service Acceptance Trials, Special Weapons Phase, which had been performed with BuNo 135413. In the conclusion of this report, NASWF reported that the A3D-1 was a satisfactory aircraft for carrying and delivering the Marks 5, 6, 7, 8, 12, 15 Mod 0, 18, and 91 special weapons, and recommended that it be accepted as a carrier for these special weapons.

March 1957: To publicize the capabilities of its new heavy attack aircraft, the Navy initiated planning for transcontinental speed run attempts as early as November 29, 1955, during a conference at NATC when representatives from NATC, CNO, BuAer, and Fleet Weather Central briefed pilots on the best methods of flying the jet stream. Unfavorable wind conditions resulted in unsuccessful attempts being made on January 31 and March 21, 1956, and further attempts were postponed until the following winter. At last, the Navy's efforts were rewarded on March 21, 1957, when BuNo 135431, flown by Cdr. Dale W. Cox, Jr. (the project pilot for *Project AROWA*, the Navy's research program on jet stream flying), Lt. Russell H. Baum and Marine TSgt. Robert L. Butts, set two transcontinental records.

Taking off from Los Angeles International Airport, the aircraft flew to Floyd Bennett Field, New York, covering 2,445.9 miles (3,935.5 km.) in 3 hours 59 minutes 15.2 seconds, but failed to break the West-East record set two years earlier by an Air Force Republic F-84F. After a 21 minute refueling stop, the same crew flew the aircraft back to LAX in 5 hours 14 minutes 57.6 seconds, thus breaking the East-West record set 22 months earlier by a California ANG North American F-86A. The total round-trip elapsed time of 9 hours 35 minutes 48 seconds also bettered the record set by the California ANG fighter in May, 1955.

July 1957: The SR-38 Part II demonstration was completed.

The challenge of designing a carrier-based strategic nuclear bomber had been successfully met by Ed Heinemann and his El Segundo design team.

BuNo 135418 is resplendent in its high-visibility scheme (orange dayglo nose, rear fuselage, tail surfaces, and wing tip panels) on display at NAS Moffett Field, California. Note the early-style pointed nose radome and the absence of tail guns.

Also finished in the high-visibility scheme—a far cry from today's tactical scheme designed to reduce visibility and infrared detectability—BuNo 135414 evidenced the then overriding concern of preventing mid-air collisions. The forward fuselage panel around the twin ATM exhausts was left in the standard gray paint.

Chapter 4:
The Bomber Variants

BuNo 135421 was one of 38 A3D-1s built under Contract 55-190C, the second "Skywarrior" production contract. It is seen here late during its operational life, by which time it had been redesignated A-3A, prior to its transfer to NAS Lakehurst, New Jersey, for use as a maintenance trainer. The aircraft has been fitted with a replacement intake nose ring and thus displays a double danger chevron on the intake of its right Pratt & Whitney J57-P-6B turbojet.

Only 214 Skywarriors were built as A3D-1 and A3D-2 bombers. However, six of the A3D-1's were almost immediately converted (one becoming the YA3D-1P photographic reconnaissance prototype and five being modified into YA3D-1Q/A3D-1Q electronic reconnaissance aircraft) while 90 A3D-2's were later converted into KA-3B tankers and EKA-3B tanker and countermeasures strike support aircraft. All aircraft built as bombers retained the fuselage design developed for the XA3D-1. In addition, the same basic fuselage design was to have been used by a more advanced bomber variant, the still-born A3D-3, and by several proposed variants intended to fulfill a variety of roles.

The A3D-1, the First Production Model

The first 12 A3D-1's (BuNos 130352 to 130363) were the subject of a Letter of Intent issued on February 10, 1951, some 20 months prior to the XA3D-1's maiden flight. When ordered, these aircraft were to be powered by two 7,500 lb. (3,402 kg.) thrust Westinghouse J40-WE-12 turbojets. However, as related earlier, a change order providing for the substitution of J57's for the unsatisfactory J40's was later issued and incorporated in the definitive contract, NO(a)s 51-632 dated October 22, 1952. A follow-on order for 38 A3D-1's (BuNos 135407 to 135444) was covered by Contract NO(a)s 52-981 (renumbered NO(a)s 55-190c Lot I after a number of changes were incorporated).

Changes incorporated in the second A3D-1 production batch included some dictated by flight trials (e.g., modification of the fuel transfer and dumping systems, elevator trim actuator fix, and substitution of improved windshield wipers effective up to 300 knots/555 kmh in place of the initial sets which became ineffective at speeds above 125 knots/230 kmh) and others to improve operational flexibility (e.g., enlargement of the bomb bay to carry four instead of three 2000-lb./907-kg. M66 GP bombs or Mk 25 mines).

The first A3D-1's were initially powered by J57-P-1's with J57-P-6A's being introduced during trials and the specified J57-P-6B's being fitted as soon as sufficient engines were available. Until the P-6B's were installed, the Skywarriors did not have the required dual bleed system. Unfortunately, production delays at Pratt & Whitney, which resulted from the very large number of J57's then being ordered for Air Force and Navy aircraft, resulted in an extremely marginal supply of J57-P-6B's for the A3D-1's until September, 1956. Accordingly, to prevent delay in scheduled production deliveries of A3D-1's, the Navy elected to have early Fleet deliveries made with a J57-P-6A on the starboard side and a J57-P-6B on the port side. This installation enabled the use of improved duct bleed for reduction of cockpit contamination (a problem which had plagued the A3D-1 since its first flight) as the port engine supplied the air.

All A3D-1's were delivered with the pointed radome and tail turret and, as far as it has been feasible to ascertain, none were retrofitted with the later flat radome or dovetail ("duckbutt") fairing.

Photographed in Dayton, Ohio, in September, 1954, BuNo 130356 is typical of early production A3D-1s with their glossy sea blue finish, fin tip fairing, and Westinghouse Aero 21B tail turret. Note the different angles at which the nozzles of the JATO bottles were mounted.

The old and the new paint schemes applied to A3D-1s during production are displayed by these two factory photographs. BuNo 135422, seen on the left flying over Long Beach harbor, is in the original glossy sea blue finish which had been adopted for carrier-based aircraft in June, 1944, when Specification SR-2e was issued. BuNo 138905, which was photographed on January 4, 1957, bears the new light gull gray and insignia white scheme complying with Mil-C-18263(Aer) which was issued in February, 1955.

With the exception of BuNo 130353, which was experimentally fitted with a refueling probe on the port side for use as a receiver during tanker trials, none of the A3D-1's were fitted with either the refueling probe or the tanker package.

In service, the electronic systems fitted to most A3D-1's included AN/ARA-25 UHF direction finder, AN/ARN-14E or AN/ARN-21 VOR homing, AN/ARC-27A UHF transmitter-receiver, AN/ART-13 HF transmitter, AN/ARR-15A HF receiver, AN/APX-6B and AN/APA-89 IFF, AN/AIC-4A interphone, AN/APN-22 radio altimeter, and AN/ASB-1A bomb director. Maximum internal bomb capacity was 12,500 lb. (5,670 kg.) and an Aero 21B tail turret with two 20mm M3L cannon (500 rounds per gun) was provided for rear defense. Maximum internal fuel capacity was 4,385 gallons (16,599 liters) in two self-sealing fuselage tanks and two wing tanks.

The 50 aircraft ordered as A3D-1's were accepted between October, 1953, and October, 1956, with peak delivery occurring in May, 1956, when five aircraft were accepted. One of these aircraft became the YA3D-1P prototype, five were modified as YA3D-1Q/A3D-1Q's, and one was re-engined with J75's. Most of the remaining 43 A3D-1's saw limited service with deployable squadrons prior to being assigned to the RAG squadrons, VAH-3 and VAH-123, for use in the training role; others were used for a variety of R&D undertakings as described elsewhere in this book. Fifteen were lost in accidents and 28 were stricken or salvaged at the end of their useful lives (for details see Appendix C).

The A3D-1's were redesignated A-3A's on September 18, 1962, when the Department of Defense implemented a common designation system for all Air Force, Army and Navy aircraft; two of these aircraft used for special tests were given the status prefix letter N and became NA-3A's (BuNos 135409 and 135427). In May, 1975, two months short of the 20th anniversary of its acceptance, BuNo 135409 was salvaged at the Military Aircraft Storage & Disposition Center (MASDC), Davis-Monthan AFB, Arizona, thus ending the service life of the original A3D-1 variant.

The A3D-2, the Most Produced Model

As originally ordered, the A3D-2 was to differ from the earlier model only in having a strengthened airframe (load factor of 3.4 g versus 2.67 g for the A3D-1) and an enlarged bomb bay (the number of 500-lb./227-kg. bombs being increased from eight to 12). In addition, the A3D-2 was to receive more powerful engines (J57-P-10's with maximum and military ratings of 10,500 lb./4,763 kg. versus the 10,000 lb./4,536 kg. maximum rating and 9,500 lb./4,309 kg. military rating of the J57-P-6B's powering the A3D-1). Subsequently, however, a number of other improvements (e.g., CLE wings and tanker package—as described under the next heading, and defensive electronic countermeasures systems, DECM, fitted in place of the tail turret and comprised of the AN/ALQ-19, AN/ALQ-32, AN/ALQ-35) were incorporated during manufacture. Other improvements (e.g., provision for 748-gallon/2,831-liter auxiliary tank in the upper bomb bay area, and a revised nose radome with flat panel slanted down and aft—this "flat" nose was first fitted to improve the performance of the aircraft fitted with the AN/ASB-7 radar but was also later adopted as the new standard for *Skywarriors* with AN/ASB-1A or AN/ASB-1B radar) resulted from service changes.

Structural testing of the airframe's increased strength proved to be the source of delays when the wing failed at 120% of design load on September 12, 1955. The failed area was in the fuselage wing mating and resulted in major damage to the test hull and the need to procure a replacement static test article. The 3.4 g wing finally passed its static test in January, 1956, clearing the way for the A3D-2's first flight on June 12.

At about the same time as the A3D-2's first flight, the Navy and Douglas discussed various ways of improving the aircraft's effectiveness by replacing the unsatisfactory tail gun installation with either DECM equipment or a liquid rocket engine. A brief study of the proposed rocket motor installation revealed that it would not be suitable due to marginal gains in buffet limiting the ceiling under medium loading conditions. Accordingly, the more promising DECM installation proceeded to the detailed design phase; it was eventually incorporated in the last 21 A3D-2's during manufacture and was retrofitted to most earlier A3D-2's.

The A3D-2's service acceptance trials, including armament, electrical and electronic, service suitability, stability and control, aircraft and engine performance, and carrier suitability trials, commenced on October 17, 1957. The final report submitted by the Board of Inspection and Survey on March 30, 1959, concluded that the four aircraft assigned for trials had met or exceeded all contractual guaranteed values. The Board recommended that the A3D-2 be finally accepted for service use after satisfactory actions were taken to correct minor deficiencies (the type had entered service more than 26 months earlier when VAH-2 had taken delivery of its first A3D-2 in January, 1957).

A total of 164 A3D-2's were built, with acceptance extending from April, 1956, until January, 1961. The first 75 (BuNos 138902 to 138976) were built under Contract NO(a)s 55-190c Lot III (initially NO(a)s 54-849). They were followed by 48 identical aircraft (BuNos 142236 to 142255, 142400 to 142407, and 142630 to 142649) and 20 aircraft fitted with CLE wings, provision for the tanker package, and a refueling probe (BuNos 142650 to 142665, and 144626 to 144629) under Contract NO(a)s 55-190c Lots VIII and X. A final batch of 21 A3D-2's (BuNos 147648 to 147668), with CLE wings, tanker package, and refueling probe, was built under Contract NO(a)s 59-0150; aircraft in this batch were delivered with an AN/ASB-7 bomb director system in lieu of the AN/ASB-1A and with the DECM system in the dovetail fairing in place of the tail turret.

On delivery, late production A3D-2s typically had an electronic suit comprised of AN/ARN-21 TACAN, AN/ARC-1 VHF and AN/ARC-27A UHF transmitter-receivers, AN/ARN-14E VOR homing, AN/APN-22 radio altimeter, AN/ARA-25 and AN/ARN-6 UHF direction finder, AN/APX-6B IFF transponder, AN/ARC-38 HF transmitter-receiver, and three DECM systems (AN/ALQ-19, -32, and, -35). Normal internal fuel capacity was 5,086 gallons (19,252 liters) but a 1,224-gallon (4,633-liter) auxiliary tank could also be fitted in the lower portion of the bomb bay when the aircraft carried no bombs and was used as a tanker.

Quickly displacing the A3D-1's in service with deployable squadrons, the A3D-2's were redesignated A-3B's in September, 1962. By then, their primary mission was no longer nuclear deterrence, but the type was gaining in importance as the Navy's primary tanker. This new role eventually led the Naval Air Rework Facility at NAS Alameda, California, to convert 90 A-3B airframes to the KA-3B (85 aircraft) and EKA-3B (five aircraft) configurations, with several aircraft being changed back and forth between these two configurations. Of the remaining 74 A3D-2 airframes originally delivered by Douglas, 53 were lost as a result of accidents, 18 were stricken or salvaged, one (BuNo 138931) was still stored at MASDC at the end of 1985, and two were still active at that time as NA-3B's (BuNos 138938 at PMTC and 142630 at the NWC China Lake).

The Development of Inflight Refueling Capability

While the J57-powered A3D-1 was going

through its test and evaluation program, the Navy acquired a fair amount of experience with North American AJ *Savages* modified to serve as air refueling tankers and in 1955 decided that all its new fighter and attack aircraft would be equipped for refueling in flight. Prior to reaching this decision which was to create an immediate need for tankers, the Navy had already requested that Douglas initiate studies of a dedicated tanker version of the *Skywarrior* and of a tanker kit to endow the A3D-2 bomber with limited inflight refueling capability. Not wanting to encumber his *Skywarrior* with the necessary equipment, Ed Heinemann at first expressed his displeasure. The Navy, however, was adamant and Douglas proceeded with the required design work.

To meet the first of these requirements, Douglas designed a rigid tubular refueling device to be stowed fully within the lower section of the bomb bay and extended hydraulically downward for refueling operation. An auxiliary tank was to be fitted in the upper part of the bomb bay to increase the amount of fuel that could be transferred. Nicknamed "flying pipe," the rigid refueling device was comprised of two pipes hinged at the rear ot the bomb bay and joined together in a "V" shape; a single pipe ending with a standard refueling drogue was attached at the point of the "V" by means of a rotating joint.

Douglas fitted the flying pipe device to BuNo 138903 to demonstrate the equipment proposed for the dedicated tanker. On July 30, 1956, the first inflight evaluation of this prototype installation revealed that considerable buffet was encountered by the receiver aircraft due to airflow disturbance created by the tanker's open bomb bay. Within two weeks, the decision was reached to discontinue trials with the flying pipe system and to replace it with a "hose reel" system as used on the *Savage* and as previously proposed for the conversion kit to be provided for A3D-2 bombers.

The hose reel selected for initial installation in BuNo 138903 was the British-designed, electrically-operated A-12B-1 supplied by Flight Refueling, Inc. The reel drum was housed in the rear of the bomb bay, leaving space forward for a reduced bomb load, and the hose was extended through a ventral fairing, thus eliminating the need to open the bomb bay doors during inflight refuelings. Trials began in the fall of 1956, with BuNo 130353 (an A3D-1 which was used to test the refueling probe being developed for late production A3D-2's) and a Grumman F9F-6 (BuNo 131043) being used as receivers.

Although the A-12B-1 hose reel was more satisfactory than the flying pipe system, difficulties were still encountered by the receiver as the hose tended to twirl in the slipstream. A change to the Flight Refueling A-12B-7 hose reel with hydraulic drive and control cured the problem. Most of the development flying with the definitive refueling package was made by BuNos 142651 and 142652. Following service suitability trials conducted with these two aircraft from April 15 to July 17, 1959, the tanker kit was approved for use in late production A3D-2's beginning with BuNo 142650. These aircraft were also delivered with a refueling probe bolted on the port side and feeding into the forward fuselage tank, provision for a 1,224-gallon (4,633-liter) auxiliary tank in the lower section of the bomb bay, and CLE wings.

To exploit fully the tanker capability, it was found desirable to increase the maximum catapult take-off weight of the A-3D-2. Unmodified aircraft were limited to a catapult weight of 70,000 lb. (31,751 kg.) but were launched during tests at NATC Patuxent River at weights up to a maximum of 76,500 lb. (34,700 kg.). At this high gross weight, however, handling and performance characteristics were found to be too marginal for safe operation by fleet personnel. Accordingly, Douglas engineered a localized redesign of the wings to incorporate both minor airfoil improvement and inboard slats and to increase wing area from 779 to 812 sq. ft. (72.37 to 75.44 sq. m.). In addition to allowing an increase in maximum catapult launch weight to 84,000 lb. (38,102 kg.), the new Cambered Leading Edge (CLE) wings promised to offer the following advantages: (1) increased combat ceiling, cruise altitude, and combat radius; (2) reduced wind-over-deck requirements for catapult launch and arrested landing; (3) lower stalling speed, and (4) improved stall warning by reducing the margin and intensity of buffet at the stall. The only minor drawback of the CLE wings was a 3 to 4 knot (5 to 7 kmh) reduction in top speed.

In August, 1956, satistifed that the CLE wings would indeed provide the required improvements, BuAer authorized tests with two modified aircraft (BuNos 138918 and 138932) and incorporation of CLE wings in late production A3D-2's, the last two A3D-2P's, the last 12 A3D-2Q's, and all A3D-2T's. The 84,000 lb. catapult launch weight was demonstrated by LCdr. Edward A. Decker on August 25, 1959, when he made 13 launches and 12 arrestments aboard the USS *Independence* (CVA-62). Twenty-seven years later, the A3D-2 remains the heaviest aircraft ever catapulted from a carrier. In practice, however, the A3D-2's with CLE wings were limited to maximum take-off weights of 73,000 lb. (33,112 kg.) for carrier operation and 78,000 lb. (35,380 kg.) for land operation.

Often, but mostly outside the A-3 community, A-3B's fitted with the tanker package have been erroneously identified as KA-3B's. The KA-3B designation, however, refers only to those aircraft modified by NARF Alameda during the Southeast Asia War for use as dedicated tankers, and all photographs of *Skywarrior* tankers taken prior to the spring of 1967 can be safely identified as photographs of A-3B's, not of KA-3B's. When fitted with the tanker package, A3D-2's (later A-3B's) retained their original designations and were known as "pathfinders." At first these aircraft were normally operated without the tanker package, the package being installed in approximately four hours when needed either to extend the range of *Skywarrior* or other naval aircraft, thus providing them with the capability of reaching distant targets, or to refuel other aircraft during long delivery or positioning flights. Thus, although they were increasingly flown with their tanker pacakge installed, late production A3D-2's were primarily operated in their originally intended role of heavy attack. The KA-3B's, on the other hand, were exclusively used as tankers and had the hose reel permanently installed.

The KA-3B and EKA-3B, the Vietnam War's Variants

Soon after U.S. naval aircraft began flying combat sorties over North Vietnam and Laos on a regular basis, the Heavy Attack Squadrons started using their A-3's almost exclusively in the tanker role, not only to extend the range of combat aircraft assigned distant targets but more importantly to provide unscheduled refueling for aircraft returning with battle damage or short of fuel after air combat operation. That the latter type of air refueling activities rapidly grew in importance is evidenced by the fact that between the beginning of 1956 and the summer of 1967 A-3B's were credited for saving some 380 Navy carrier-based aircraft valued at over $450 million.

While A-3B's fitted with tanker packages were doing an outstanding job during the early years of the Southeast Asia War, it was becoming evident that the effectiveness of these aircraft could be further improved if all their no longer required bombing equipment was removed and their weight reduced to increase the amount of transfer fuel available for any given launch weight. At the same time, the Navy also developed another urgent requirement for an aircraft to replace its only carrier-based electronic countermeasures aircraft, the obsolete piston-powered Douglas EA-1F *Skyraider*.

As the *Skywarrior* had the necessary weight lifting capability and internal volume to fulfill both the dedicated tanker and electronic countermeasures missions, the Navy decided to have the A-3B's modified while they underwent overhaul and repair at NARF Alameda. The modification program, which peaked in 1967-68, resulted in the delivery of two distinct groups of aircraft: (1) 85 A-3B's were reworked as dedicated tankers and designated KA-3B's on April 7, 1967, and (2) five 3-B's were redelivered as Tactical and Countermeasures Strike Support Aircraft (TACOS) and designated EKA-3B on February 16, 1967. Later permutations resulted in 34 KA-3B's being brought up to EKA-3B standard while three EKA-3B's became KA-3B's prior to rejoining the Fleet. After being replaced in VAQ squadrons by Grumman EA-6B *Prowlers*, the surviving EKA-3B's had their ECM equipment removed to end their useful life as KA-3B's.

By the end of 1985, the fate of the 90 aircraft having borne the KA-3B and EKA-3B designationals was as follows: 16 had been lost in accidents, 24 had been stricken or salvaged, 35 were stored, 13 (two KA-3B's and 11 EKA-3B's converted back to KA-3B standard) were still active, and two were being refurbished by NARF Alameda.

After they had been modified by NARF Alameda, the KA-3B's differed from the A-3B's primarily in having the bomb bay spoiler and DECM equipment removed, the bomb bay modified so that the doors could only be operated on the ground, and the A-12B-7 hose reel (1) and

---

[1] As the A-12B-7 hose reel was no longer in production, a reels' shortage was solved by retrieving from MASDC 40 electrically-powered A-12B-1 hose reels which had become surplus when Boeing KB-50J's were phased out by the Air Force. Ten of these units were cannibalized for parts while 30 were rebuilt to the A-12B-7 standard for use in KA-3B's and EKA-3B's.

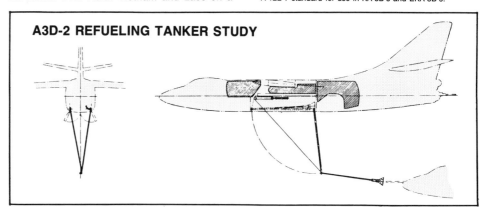

A3D-2 REFUELING TANKER STUDY

BuNo 138903, an A3D-2 fitted with an electrically-operated A-12B-1 hose reel unit, is seen during refueling tests utilizing BuNo 130353, the second A3D-1. The latter had been fitted during the fall of 1956 with a refueling probe that was then under development for late production A3D-2s.

The prototype tanker (A3D-2 BuNo 138903) and receiver (A3D-1 BuNo 130353) flying formation during tests in late 1956. Note that the refueling probe fitted to this A3D-1 penetrated the fuselage below the cockpit whereas that adopted for late production A3D-2 extended externally much more aft.

Still bearing the markings of VAH-2 with which it had returned from a war cruise in the Gulf of Tonkin prior to its conversion to the EKA-3B configuration, BuNo 147655 displays the EKA-3B's characteristic fuselage side blisters, ventral canoe, and refueling hose fairing.

BuNo 130355 photographed at O&R Jacksonville, Florida, in January, 1959, after installation of the engine pods manufactured by Ludwig Honald, Inc. and housing Pratt & Whitney J75-P-2 turbojets. This version of the J75, intended for the Martin P6M-2 flying-boat, had a maximum thrust rating of 15,800 lb.

The J75-powered A3D-1 is shown at NATC Patuxent River on April 10, 1959. The nacelles were similar to those fitted to DC-8 Series 20 and 30 aircraft but the experimental A3D-1 proved of little value to Douglas as the first JT4A-powered DC-8, a Series 30 for Pan American, had first flown on February 20, 1959.

associated equipment permanently fitted.

Three different types of KA-3B's were obtained. Type I and Type II aircraft used late production airframes with CLE wings, Liquidometer fuel quantity indicating system, and a six-step fuel balance system using forward-to-aft fuel transfer pumps. Type III aircraft were based on older airframes with standard wings and Avien fuel balance and quantity indicating systems using gravity transfer from forward to aft fuel tank. All three types had tanks with a total volume of 5,282 gallons (19,994 liters), but usable fuel was 5,086 gallons (19,252 liters) for Types I and II aircraft versus 5,169 gallons (19,566 liters) for Type III aircraft. Differences between Type I and Type II aircraft were limited to instrument and console panels.

Easily distinguishable from other *Skywarriors* as they were fitted with large blisters fore and aft of the wings on both sides of the fuselage, the EKA-3B's were dual role aircraft. In addition to the equipment and systems common with those fitted to the KA-3B's, they were equipped with an AN/APN-153(V) doppler radar system, an AN/ASN-66 dead reckoning computer, and a comprehensive ECM fit. The EKA-3B's ECM equipment included two AN/ALT-27 S-band barrage or spot jammers with antennas in a ventral fairing; an AN/ALQ-92 VHF radar and voice communications jammer with antenna and equipment filling the side blisters; an AN/ALR-28 X-band direction finding receiver with its antenna in a small blister atop the fin; an AN/ALR-29 L and C-band panoramic receiver; an AN/ALR-30 S-band panoramic receiver; an AN/APR-32 SAM warning receiver; two AN/ALQ-41 and two AN/ALQ-51 DECM systems; an AN/ALA-3 or AN/ULA-2 L, S and C-band pulse analyzer[2]; a C-7317/AL DF indicator[3]; an AN/APA-69 L, S and C-band DF antenna in the ventral fairing, aft of the AN/ALT-27 antennas and ahead of the refueling drogue; and two AN/ALE-2(M) chaff dispensers. [The ALR-29 and ALR-30 interfaced with the ALT-27; the ALR-29 interface may not have been functional until after Vietnam.] In spite of the complexity of operation of its ECM equipment, the EKA-3B still had a crew of three, with the third crew member being primarily responsible for the monitoring and use of the ECM equipment.

The A3D-3 and Other Re-engined Variants

Not resting on its laurels after securing orders for the A3D-2, Douglas submitted preliminary details for the A3D-3 model in 1955. Many changes were proposed, but the most likely to be incorporated in a production version included various minor wing improvements, all-flying horizontal tail surfaces, a rudder pedal nose wheel steering system, and ECM equipment in the tail cone in place of the twin-cannon turret of earlier models. Likewise, several engine types were evaluated with proposed A3D-3 versions including aircraft powered by either late model J57's or the 50% more powerful Pratt & Whitney J75's and Wright J67's. However, after showing some interest in these proposals, the Navy elected to procure additional A3D-2's rather than fund the development of yet another *Skywarrior* model as the use of more powerful engines only resulted in nominally improved level and climb speeds while reducing combat radius. Furthermore, the supersonic North American A3J-1 *Vigilante* had been ordered and the submarine-launched *Polaris* missile was under development to take over the *Skywarrior's* nuclear deterrence task. Accordingly, all work on the A3D-3 was terminated in 1956.

Earlier, Douglas had also studied the feasibility of replacing the J57's of the *Skywarrior* with

---
[2] For analysis of signals received by ALR-29 and ALR-30.
[3] C-7317/AL indicator used to display bearings of signals received by ALR-29 to ALR-30 and to optionally position ALT-27 steerable jamming antennas.

4,000 shp Allison T56 turboprops mounted in nacelles attached to standard pylons. Even when using large diameter four-blade propellers, sufficient tip clearance was provided. Calculated performance indicated that the T56-powered *Skywarrior* would be only slightly slower than the standard aircraft but would show a fairly sizeable increase in range. In other ways, however, the switch to turboprops was considered a step backward and this project was also allowed to lapse prior to reaching the hardware stage.

As previously indicated, Douglas had considered using Pratt & Whitney J75's to power the proposed A3D-3 model. However, on July 25, 1955, engineering problems, time considerations, and only marginal advantages over the use of late model J57's prompted the Navy to reject this powerplant installation for production versions of the *Skywarrior*. Nevertheless, the Navy wished to acquire inflight experience with this powerplant which was to power the Martin P6M-2, and Douglas was keen to obtain similar experience as, in its JT4A version, the new and more powerful Pratt & Whitney turbojet had already been selected to power the Series 20 and 30 of its DC-8 jetliner. With the desirability of flight testing the J75 satisfying the requirements of both parties, an agreement was reached whereby an A3D-1 would be re-engined with this powerplant and used to accumulate 2,000 flight hours of flight time in 1½ years, from January, 1959, through June, 1960. The aircraft selected as the testbed was BuNo 130355.

Preliminary design work on the J75 installation was made by Douglas and Pratt & Whitney Aircraft did the final engine installation design with some assistance from Douglas. The engine pods were manufactured by Ludwig Honald, Inc. of Philadelphia, Pennsylvania, and Douglas supplied some airplane modification kits. The pods were shipped from Philadelphia to O&R Jacksonville, Florida, where Pratt & Whitney and Navy personnel installed the new powerplant and test instrumentation.

To ensure compliance of the installed engines with contractual performance requirements, Pratt & Whitney was assigned responsibility for initial flight trials. Accordingly, a P&W engineering flight test pilot, Paul D. Berk, first flew the J75-powered A3D-1 at NAS Jacksonville on January 23, 1959. Three more flights were made at this station and on February 11, Paul Berk ferried the aircraft to NAS Patuxent River, Maryland. In four more flights, Pratt & Whitney completed verification of engine guarantees at maximum flight speeds and during climbs to altitude and investigated flutter characteristics of the A3D wing with J75 nacelles. Stall characteristics and stall recovery were explored during whip turns.

After completing 13 hours 35 minutes of testing, Pratt & Whitney handed over the J75-powered aircraft to the Naval Air Test Center where Navy pilots were to conduct accelerated service testing. Some delays were incurred when the oil cooler air inlet duct, located slightly aft and below the main air intake, failed and had to be replaced by a revised design. By then, however, the trial program had lost its raison d'être as on August 31, 1959—less than ten months after the termination of the LTV F8U-3, the only other naval aircraft powered by the J75—the Navy cancelled the production contract for the Martin P6M-2, thus eliminating the need for further testing of the J75-powered *Skywarrior*. Thereafter BuNo 130355 remained on limited flight status at NATC Patuxent River until August, 1964, when it was flown to Litchfield Park, Arizona, to be placed in storage. Stricken from inventory in May, 1965, the aircraft had been flown for a total of 926 hours, including 677 hours since it had been re-engined with J75's.

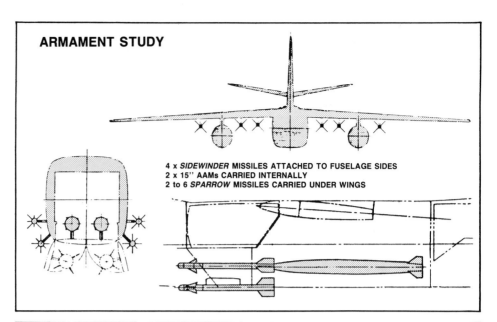

ARMAMENT STUDY

4 x *SIDEWINDER* MISSILES ATTACHED TO FUSELAGE SIDES
2 x 15" AAMs CARRIED INTERNALLY
2 to 6 *SPARROW* MISSILES CARRIED UNDER WINGS

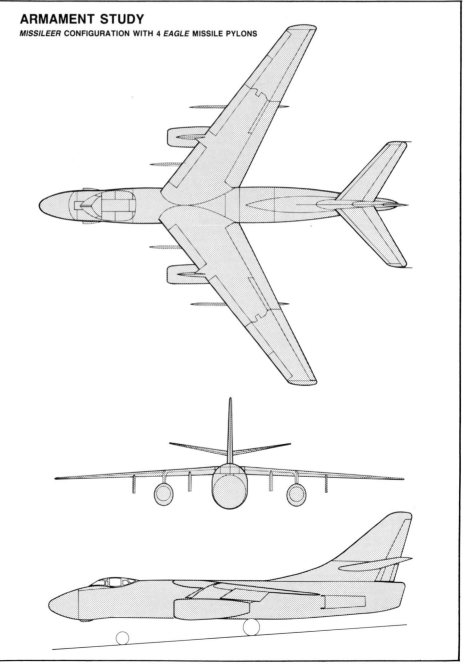

ARMAMENT STUDY
MISSILEER CONFIGURATION WITH 4 *EAGLE* MISSILE PYLONS

Skywarrior Production Summary
The Bomber Variants

	Contract NO(a)s	FSN	BuNos	No. of aircraft	Remarks
XA3D-1	10414	7588/7589	125412/125413	2	
A3D-1	51-632	9253/9264	130352/130363	12	(1)
	55-190C	10300/10337	135407/135444	38	
A3D-2	55-190c	10763/10837	138902/138976	75	
	55-190c	11562/11581	142236/142255	20	
	55-190c	11584/11591	142400/142407	8	
	55-190c	11693/11712	142630/142649	20	
	55-190c	11713/11728	142650/142665	16	(2)
	55-190c	12024/12027	144626/144629	4	(2)
	55-0150	12412/12432	147648/147668	21	(2)

Remarks:
(1) See YA3D-1P and YA3D-1Qs as listed in Chapter V.
(2) With CLE wings and tanker package.

While on the subject of possible re-engining programs for the *Skywarrior*, mention must be made of the contemplated installation of secondhand Pratt & Whitney JT8D turbofans. In 1981-82, when it had become obvious that the A-3 was likely to soldier on into the early 1990's, NARF Alameda studied the feasibility of replacing the J57 turbojets, for which spares were getting increasingly difficult to find, with a more modern powerplant. Taking its inspiration from the Air Force program for the re-engining of Boeing KC-135A's with JT3D turbofans taken from surplus Boeing 707-120B commercial transports, NARF Alameda proposed to fit JT8D's from surplus DC-9 Series 10 jetliners. The idea appeared technically feasible as the pylon installation of the *Skywarrior*'s engines would have required only a relatively simple redesign of the nacelles to fit JT8D's. However, the discovery in the Navy's supply pipeline of a number of low-time J57-P-10 turbojets—the standard powerplant of the *Skywarriors* remaining in service—eliminated the need for this expensive re-engining.

The Proposed Fighter, Interdiction/Ground Attack, and Target Towing Versions

As early as the summer and fall of 1954, Douglas proposed several unusual configurations to handle missions not initially foreseen for the *Skywarrior*. One of these configurations envisioned the carriage of four *Sidewinder* air-to-air missiles attached to the sides of the fuselage on special Douglas launching racks and endowing the heavy attack version with self-protection capability. A "fighter" version was proposed with an armament consisting of up to six *Sparrow* missiles carried beneath the wings (one outboard and two inboard of the engine nacelle on each side). With six missiles, changes in performance were calculated to be minus 11 knots (20 kmh) in maximum speed at 35,000 ft. (10,670 m.), minus 1,600 ft. (490 m.) in combat ceiling, and minus 160 nautical miles (295 km.) in combat radius.

For the air support role, Douglas proposed the installation of 120 two-inch (50.8 mm) folding fin rockets in the upper fuselage area and 20 five-inch (127 mm) *Zuni* rockets in the lower area. As the launch tubes for the FFAR's were angled 17° above the line of flight and those for the *Zuni* rockets 10° below the same line, close air support accuracy might have proven somewhat elusory, if not outright dangerous, for friendly "grunts"! A more practical solution was proposed for an interdiction version which was to have been armed with six forward-firing Mk-12 20 mm cannon mounted in a special fairing beneath the bomb bay with ammunition boxes (200 rpg) carried within the bay.

Douglas also proposed adapting the A3D-2 for towing targets through the use of a kit fitting in the bomb bay. The bay doors were to be removed and a floor substituted on which would be installed the target winch and controls, with space provided for storage of two target sleeves. Access to this compartment, in which was included accommodation for the winch operator, was to have been through the existing companionway. A retracting cable guide was provided to permit access to the tow target cable for attaching targets.

None of these proposed variants was proceeded with.

The Douglas Model D-790 Missileer

Among the many uses which were considered for the *Skywarrior*, none was more unconventional than that envisioned for the Douglas Model D-790 *Missileer*, a proposed standoff fighter for use in the defense of the U.S. Fleet and the North American continent. Preliminary design work for this model was undertaken by Douglas in response to Proposal Request No. AER.24770 issued by the Bureau of Naval Weapons, with a Performance Data and Flying Qualities Report being issued by the manufacturer on February 19, 1960. Basically retaining the standard airframe of late production A3D-2's, with cambered wing leading edge and "dovetail" rear fuselage, the D-790 was to have been fitted with a powerful AN/APN-122(V) radar in a bulbous nose. Its sole armament was to have consisted of Bendix/Grumman AAM-N-10 *Eagle* long-range air-to-air missiles, four *Eagles* being normally carried in the slightly modified bomb bay of the *Skywarrior*. Overload conditions provided for the carriage of four additional missiles beneath the wings, one on each side of the engine nacelles.

Initially Douglas contemplated using a pair of 12,000 lb. (5,443 kg.) thrust Pratt & Whitney JT3C-14 turbojets, but before submission of the proposal a switch was made to two General Electric J79-MJ98C turbojets rated at 11,650 lb. (5,284 kg.) on take-off. With these engines, the D-790 *Missileer* was expected to have a top speed of 635 mph (1,022 kmh) at sea level in clean condition and 588 mph (946 kmh) when carrying eight AAM-N-10 missiles. With four missiles in the bomb bay and two beneath the wings, the D-790 was anticipated to be able to loiter for 4.5 hours some 150 nautical miles (280 km.) from its base or carrier, mission duration being 5.8 hours (including transit time and 10 minutes of combat at full military power). The design weight of the D-790 was 55,942 lb. (25,375 kg.) and its maximum take-off weight 79,410 lb. (36,020 kg.).

The *Missileer* concept attracted strong interest on the part of the Navy; however, rather than ordering the D-790, a new request for proposals was circulated to the industry. Douglas' response was the D-976, a new design inspired by its earlier F3D *Skyknight*, which was to be powered by two Pratt & Whitney TF30-P-2 turbofans and was to carry six AAM-N-10 *Eagle* missiles on wing racks. On December 15, 1960, Douglas was awarded a contract for the D-976, which was to be produced as the F6D-1, but the Navy lost interest in the *Missileer* concept and cancelled the F6D-1 contract on April 25, 1961.

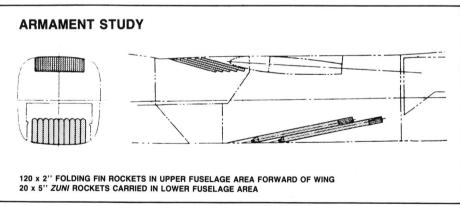

ARMAMENT STUDY
120 x 2" FOLDING FIN ROCKETS IN UPPER FUSELAGE AREA FORWARD OF WING
20 x 5" ZUNI ROCKETS CARRIED IN LOWER FUSELAGE AREA

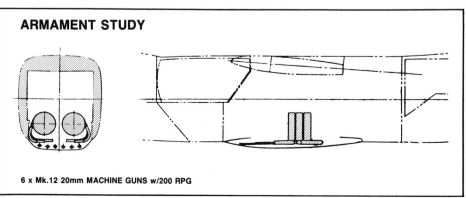

ARMAMENT STUDY
6 x Mk.12 20mm MACHINE GUNS w/200 RPG

Chapter 5: The Versions

The YA3D-2P, BuNo 142256, photographed at the El Segundo Division on March 25, 1958, was structurally complete, but nearly six months were to elapse before its systems were installed and flight trails could commence on September 9, 1958. Modified as an NRA-3B in 1963, this aircraft has had a long and distinguished career in various test configurations. In early 1987, as this book went to press, BuNo 142256 was still bailed to Westinghouse.

With the exception of the sole A3D-1P and the five A3D-1Q's which were obtained by modifying A3D-1 airframes, the *Skywarrior* variants collectively known as the "versions" in the A-3 community all used new airframes which differed markedly from those of the bomber variants. Actually, the changes between the A3D-2 airframe and that developed for the versions under the unofficial designation "A3D-2B" were so significant and numerous that it is surprising a new aircraft configuration number was not given to the versions. Indeed, it would have been more appropriate to call them A3D-3P, A3D-3Q and A3D-3T instead of retaining the -2P, 2Q and 2T designations which implied that they were mere adaptations of the A3D-2 for the photographic reconnaissance, electronic reconnaissance, and training roles.

The common "A3D-2B" airframe developed for the versions retained the 3.4 g wings of the A3D-2 (or, for late production aircraft, the CLE wings), but featured a revised fuselage with a pressurized compartment in place of the forward fuel cell and front half of the bomb bay. The entire crew accommodation, including the cockpit and fuselage compartment, was pressurized to 7.5 psi differential over flight pressure versus 3.3 psi for the bomber variants. The increased pressurization necessitated the use of a stronger canopy (with heavier framing and thicker glazed area), the elimination of the bomb bay doors, and changes to the escape chute. The inner door was dispensed with to provide open access between the cockpit and the fuselage compartment, the thickness of the outer door was increased to seal off the pressurized area, and a small "blow-down" door was incorporated into the outer door for depressurization of the cabin prior to bail out.

To provide safe escape for their larger crews, the A3D-2Q's and A3D-2T's were fitted with a square hatch over the crew compartment (in addition to the roof hatch in the cockpit which was reduced in size and located further aft than in the basic A3D-2) and a hatch on the starboard side of the fuselage which was provided for emergency egress only. The A3D-2P only had the smaller cockpit hatch. The larger crew of the versions also necessitated replacing the two-bottle, 20-liter oxygen system of the bombers with two 10-liter LOX (liquid oxygen) converters.

In the versions, the auxiliary power system was revised with the two air turbine motors (ATM's) being relocated side-by-side beneath the cockpit instead of in tandem on the port side as installed in the bombers. This modification could be easily spotted externally, as the bombers had twin round ATM exhausts on the port side whereas the versions had one square exhaust port on each side. In addition, emergency power, which in the bombers was provided by batteries, was generated for the versions by a retractable ram air turbine (RAT) located on a small door below the starboard ATM exhaust. Less significant differences included radar optimization for search and navigation purposes (the radar set fitted to the A3D-2P and A3D-2Q versions being then designated AN/ASB-1B; the A3D-2T initially retained the AN/ASB-1A set) and the repositioning of the Goodrich multiple disc wheel brakes outboard of the wheel hubs rather than concentrically inside the rims as in the bomber variants (the wheel

The YA3D-2Q, BuNo 142257, was the 149th "Skywarrior" to be built. It is seen here on a taxiway at the Los Angeles International Airport during manufacturer's trials. The aircraft was later brought up to full production standard and was lost in 1974 while serving with VQ-2.

Redesignated NRA-3A when it was photographed at the Naval Air Development Center, Johnsville, Pennsylvania, BuNo 130358 had been built as an A3D-1 and was later fitted with a photographic reconnaissance kit to become the sole YA3D-1P. Unlike the A3D-2P version, this prototype retained the standard bomber airframe.

For Project "PRESS" (Pacific Range Electronic Signature Studies) BuNo 142256, the YA3D-2P prototype, was modified as a flying laboratory to track and observe re-entry vehicles. For that purpose, optical and infrared equipment were installed in a Boeing B-50 turret mounted aft of the cockpit.

This A3D-2P, BuNo 144828, in standard initial configuration with pointed radome and tail turret, was photographed at NAS Alameda on October 28, 1960 prior to application of squadron markings. This aircraft was lost while flying from NAS Cubi Point on June 16, 1967 while it was assigned to VAP-61.

BuNo 144849, the sixth A3D-2Q, while on display at NAS Alameda during May, 1960. The single ATM exhaust port on the left side of the forward fuselage is typical of this version. During early 1987, the aircraft was still operated by VQ-1 out of NAS Agana, Guam.

hubs of the versions protruded about 3 inches/7.5 cm. whereas those of the bombers were flush).

The A3D-1P and A3D-2P, the Photographic Reconnaissance Versions

The Navy first expressed interest in a photographic reconnaissance version of the *Skywarrior* in September, 1952, when it instructed Douglas to investigate the feasibility of a photo kit for installation in the A3D-1 bomb bay. Design work on this kit subsequently revealed that dedicated photographic (P) and electronic (Q) reconnaissance versions would have substantially greater capabilities than offered by A3D-1's fitted with reconnaissance kits. Accordingly, only one photo kit was ordered for installation in the seventh A3D-1 airframe, BuNo 130358.

Redesignated YA3D-1P (with the Y prefix being later dropped from its designation), the prototype of the photographic reconnaissance version had its kit installed at the factory prior to completion. Basically, the aircraft retained most features of the standard A3D-1 bomber (notably its three-man crew) but had its photo kit fitted in the bomb bay with the doors being replaced by metal skin and access panels. The forward portion of the kit contained up to seven cameras including forward oblique (K-18 or A-10), tri-metrogon (K-17, T-11 or CA-8), and vertical or split vertical cameras (K-37, K-38, K-17C or T-11). The aft section was divided horizontally in two, the upper part housing a 440-gallon (1,666 liter) fuel tank and the lower part containing either photoflash bombs (12 M-120's) or photoflash cartridges (208 M-112's or 80 M-123's) for use during night photographic missions. The forward portion of the bay was provided with a view finder, seat and camera control. The operator seat was accessible in flight through the companionway provided in the bomber variant to gain access to the special stores but was occupied only if necessary while cameras were in use. When the operator occupied this seat, the crew had to wear oxygen masks as the camera compartment of the YA3D-1P was not pressurized. At all other times, the operator, who also acted as gunners, sat on the third seat in the pressurized cockpit where he had a full set of camera selection and operating controls. The pilot was also provided with a partial set of camera switches.

In January, 1955, eight months after the YA3D-1P had entered flight trials, CNO, Photo Division, and BuAer Class Desk personnel reviewed the A3D-1P program. Although the support of a solitary aircraft of this type imposed logistic problems, the conferees agreed that its continued operation was warranted as the Photo Division had no other aircraft in this class to carry on much development work. Later, when A3D-2Ps were available in sufficient number to equip VAP-61, VAP-62, and VCP-63 and to take over photographic development work, the aircraft was redesignated NRA-3A and transferred to NADC Johnsville, Pennsylvania, for unrelated RDT&E work. Finally, the first photographic reconnaissance *Skywarrior* was salvaged in December, 1964.

First flown on September 9, 1958, with the YA3D-2P designation, BuNo 142256 was the prototype of the *Skywarrior's* dedicated reconnaissance version which had been ordered in preference to procurement of additional photo kits. When it began manufacturer's flight trials at Edwards AFB, this aircraft had not yet been fitted with a refueling probe on the port side, but the probe was added prior to commencement of BIS trials at NATC Patuxent River. Following satisfactory completion of these BIS trials, for which BuNo 144825 was used in the carrier suitability phase, BuNo 142256 had a long and distinguished career. In 1963, Douglas modified it into a flying laboratory to track and observe re-entry vehicles. Redesignated NRA-3B, BuNo 142256 was then used as part of Project PRESS (Pacific Range Electronic Signature Studies, a ballistic missile measurement program sponsored by DoD's Advanced Research Projects Agency). Assigned to NAPOG (Naval Airborne Project PRESS Operations Group), the aircraft was flown by a Navy crew, operated from an Air Force base (Hickam AFB, Hawaii), and staged missions from Army and Air Force facilities (the Kwajalein Army Test Site and the Eniwetok Air Force Auxiliary Airfield in the Marshall Islands). Nicknamed *Lanimoo* (Heavenly Cow in Hawaiian), the Project PRESS aircraft was fitted with optical and infrared equipment and a Mark 8 gunsight. Most of this equipment was housed in an Air Force surplus gun turret removed from a Boeing B-50.

The turret, mounted atop the fuselage just aft of the cockpit, was elevated and rotated primarily by ground control to sight and track ballistic missiles as they re-entered the atmosphere. Sensors within the turret received and transmitted infrared, visual, and ultraviolet data to recorders in the instrument rack located in the pressurized fuselage compartment. A wide-screen television gave a visual presentation of the re-entry event to the measurement officer who sat in the fuselage compartment. For ground servicing of the turret equipment, a removable platform with guard rails was provided; as the aircraft was based in Hawaii, this odd looking installation quickly became known as the "rooftop lanai."

After being used for some six years as a flying laboratory, BuNo 142256 was modified by NARF Alameda for Navy experiemnts to develop improved sonobuoys. For that purpose, large fairings were added on both sides of the lower fuselage. Still retaining these fairings, the aircraft was later bailed to Westinghouse Electric Corporation for use as a flying testbed by this company's Defense Group. At the end of 1985, BuNo 142256 was still operated by Westinghouse for avionics and ECM development work.

The A3D-2P, which was redesignated RA-3B in September, 1962, retained the three-seat accommodation of the bomber variants. Its crew was composed of the pilot, the photo-navigator/assistant pilot, and the photo-technician/gunner (guns, initially installed in the first 28 aircraft, were later replaced by DECM equipment as retrofitted to the A3D-2s). The pilot was provided with a viewfinder to fly a precise photographic course and sight the forward oblique camera. He also had controls to operate the camera selected by the photo-navigator. The latter could select and control the operation of all cameras through universal camera control equipment and a view finder. The photo-technician could change magazines (spare film magazines were carried for long missions), adjust apertures, and correct malfunctions or rearrange the cameras or stations by accessing the fully pressurized camera compartment in flight.

When entering service, and even after being supplemented by the North American RA-5C, the A3D-2P/RA-3B was one of the world's most versatile photographic reconnaissance aircraft as it was not limited to any preset or standard photographic configuration. Twelve camera mounts, 16 camera ports (beneath the fuselage and on "jowl" fairings on the lower corners of the forward fuselage), 33 camera positions, a bomb bay for flash bombs, flash cartridges, and/or electronic flash, and an associated photographic recorder permitted almost unlimited latitude in mission planning. The mount and cameras at each station were as shown on the accompanying table.

The instrumented turret of the Project "PRESS" NRA-3B, BuNo 142256, showing the platform and guardrails that were installed on the ground for servicing. This arrangement was called the "lanai on the roof" as the aircraft was based on the Hawaiian Islands.

The work platform and guard rails used by maintenance personnel to work on the NRA-3B turret were designed and fabricated at NMC Point Mugu, California, in compliance with OHSA (Occupational Health and Safety Administration) safety regulations.

NRA-3B BuNo 142256, was assigned to NAPOG (Naval Airborne Project "PRESS" Operations Group). It is seen on the ramp at Hickham AFB, Hawaii on February 28, 1964. The platform and guardrails were removed prior to flight and were usually reinstalled for maintenance shortly after the aircraft landed. Most Project "PRESS" tracking missions were flown at the Kwajalein Army Test Site and the Eniwetok Air Force Auxiliary Airfield in the Marshall Islands.

Five early production A3D-1 (BuNo 130356 and 130360 to 130363) were modified by O&R Norfolk, Virginia, into A-3D-1Q "Queer Whales" as interim electronic reconnaissance aircraft. One of these A3D-1Qs is seen here as it appeared in service with VQ-2.

STATION ONE
Mount: 189 mount with azimuth control and fixed 90 degree depression angle. IMC for the CA-13 Camera.
Cameras: CA-13, 24'' and 36'' KA-46A, 6''
CA-3-2B, 12'' KA-47A, 12''
KA-40A, 3'' KA-45A, 6''
KA-53A, 12''

STATION TWO
Mount: Special mount for the bent lens cone and motion picture camera. Fixed mount with five and seven degree depression angles.
Cameras: CA-13 Bent lens cone, 24'' and 36''
KF-8 Motion picture camera, with 2'', 4'', and 6'' lenses.

STATIONS THREE & FOUR
Mount: 5023 mount with IMC for the CA13. Rotatable to change format direction with resultant loss of IMC. Depression angles: 3°, 10.5°, 15°, 30°, and 37°.
Cameras: CA-13, 24'' and 36''

STATIONS FIVE & SEVEN
Mount: Fixed mount, 30 degree depression angle for tri-metrogon photography. Manually adjustable for pitch.
Cameras: CA-3-2B, 6'' KA-45A, 6''
CA-14, 6'' KA-46A, 6''
KA-40A, 3''

STATION SIX
Mount: 188 mount, stabilized for pitch and roll. Azimuth correction. Used as prime vertical for cartographic photography and as the vertical camera in tri-met configuration. 90 degree depression angle only.
Cameras: CA-3-2B, 6'' KA-45A, 6''
CA-14, 6'' KA-46A, 6''
KA-40A, 3''

STATIONS EIGHT, NINE, TEN, & TWELVE
Mount: 189 mount with azimuth control. IMC for the CA-13 cameras at 90 degree depression angle. 73 degree DA available as shown below for split verticals.
Cameras: 90° Depression angle
CA-3-2B, 6'' & 12'' KA-45A, 6''
CA-13, 24'' & 36'' KA-46A, 6''
K-22, 36'' KA-47A, 12''
KA-40A, 3'' KA-53A, 12''
73° Depression angle
CA-3-2B, 12'' KA-53A, 12''
KA-47A, 12''

STATIONS EIGHT-A, NINE-A, TEN-A & TWELVE-A
Mount: Same mount as used in 8, 9, 10, & 12 but rotated to an additional camera port. Only oblique photographs from this station.
Cameras: CA-13, 24'' and 36'', at 37° DA
CA-13, 24'' at 53° DA
CA-13, 36'' at 63.5° DA
K-22, 36'' at 53°, 37°, 63.5° DA
KA-45A, 6'' at 37°, 53°, 63.5° DA
KA-46A, 6'' at 37°, 53°, 63.5° DA
KA-47A, 12'' at 37°, 53°, 63.5° DA
KA-53A, 12'' at 37°, 53°, 63.5° DA

STATION ELEVEN
Mount: 187 mount, stabilized for pitch and roll. Azimuth correction. Used as prime vertical for cartographic photography. 90 degree depression angle.

Cameras: CA-13, 24'' & 36'' KA-45A, 6''
CA-3-2B, 6'' & 12'' KA-46A, 6''
CA-14, 6'' KA-47A, 12''
K-22, 36'' KA-53A, 12''
KA-40A, 3''

Note: Stabilization locked out on all cameras with focal lengths of 12'' or longer.

In the mid-sixties, when operations in Southeast Asia rendered this capability of special importance, the RA-3B was the only aircraft able to take night photographs employing M-120 or M-122 flash bombs, M-112 or M-123 flash cartridges, electronic flash, or a combination of any two of these illuminants. Flash bombs were employed at altitudes varying from 7,000 to 12,000 ft. (2,135 to 3,660 m.) and flash cartridges from 2,000 to 5,000 ft. (610 to 1,525 m.) AGL, but the electronic flash could not be used when flying higher than 2,000 ft. (610 m.) above normal terrain or 1,000 ft. (305 m.) above jungle terrain. For night operations, KA-46A, KA-47A and/or KA-35A cameras were used with 80 ASA Plus-X black-and-white film (faster film being used more rarely), Aerial Ektachrome color film, or Ektachrome Infrared (for camouflage detection).

Thirty A3D-2Ps were procured under two contracts (BuNos 142256, and 142666 to 142669 under NO(a)s 55-205 and 144825 to 144847, 146446 and 146447 under NO(a)s 57-181) and accepted between April, 1958, and April, 1960. All but the first were delivered with the refueling probe (the probe being fitted later to BuNo 142256). The last two aircraft came off the line with CLE wings and DECM installation in the tail, whereas the previous 28 A3D-2Ps initially had standard wings and tail guns. The DECM system was later retrofitted to most Fleet aircraft as was the flat radome.

Prior to 1985, 11 A3D-2P/RA-3Bs had been lost in accidents (six of which occurred during Southeast Asia operations) and two salvaged, but 17 were still in the inventory at the end of 1985. Of these, three NRA-3Bs and one RA-3B were operated by PMTC, one was flown by VQ-2 as a TNRA-3B, one NRA-3B was bailed to Westinghouse, one RA-3B was bailed to the U.S. Army for use by Raytheon, two were planned to be refurbished at NARF Alameda as ERA-3Bs, and eight ERA-3Bs were in service with VAQ-33 and VAQ-34. The PMTC NRA-3Bs and the ERA-3Bs are described separately later in this chapter.

The A3D-1Q and A3D-2Q, the Electronic Reconnaissance Versions

As the capacious fuselage of the A3D rendered it well suited to the development of a replacement of both the land-based Martin P4M-1Qs and the carrier-based Douglas AD-5Qs operated in the TASES (Tactical Signal Exploitation System) role, early in the program the Navy expressed an interest in procuring an electronic reconnaissance version of the *Skywarrior*. A mock-up of the proposed A3D-2Q was inspected in September, 1954, and five aircraft (BuNos 142257, and 142670 to 142673) were ordered under Contract NO(a)s 55-205 at the same time as the first five A3D-2Ps. Subsequently, 20 additional A3D-2Q's (BuNos 144848 to 144855, and 146448 to 146459) were ordered under NO(a)s 57-181.

While detailed engineering and systems development for the A3D-2Q were proceeding, the Navy decided to have five A3D-1 airframes (BuNos 130356, and 130360 to 130363) modified, after delivery, into interim electronic reconnaissance aircraft. Conversion of these airframes into the YA3D-1Q and four A3D-1Qs was entrusted to O&R Norfolk, Virginia, with the work being performed in 1955. The YA3D-1Q first flew in May, 1955, at NAS Norfolk and two of the "Queer Whales" were used in the A3D-1 FIP.

Retaining the fuselage bay of the bombers, the -1Qs had a crew of seven, with four operators accommodated in the unpressurized bay. ECM antennas were housed in oblong fuselage blisters (one on each side of the forward fuselage) and in short ventral canoe. In addition to their specialized ECM equipment, these aircraft were fitted with KD 2 and 3 cameras. After being brought up to Fleet delivery configuration at the end of their trials, the five A3D-1Qs (the Y prefix having by then been dropped from the designation of the first aircraft) were operated by VQ-1 and VQ-2. Three were lost as a result of operational accidents, one later became an NEA-3A while operated by PMTC at NAS Point Mugu, California, and the last was stricken in December, 1970, and eventually donated to the Pima County Musuem in Arizona.

Initially designated YA3D-2Q, the prototype of the purpose-built electronic reconnaissance *Skywarrior* first flew at El Segundo on December 10, 1958. Like the other A3D-2Qs which followed, it used the "A3D-2B" fuselage with its fully pressurized compartment, thus enabling the ECM operators to work without wearing oxygen masks as had to be done in the A3D-1Q. The first A3D-2Qs were built with standard wings and gun turret, but the last 12 had CLE wings and DECM in the dove tail fitted during production. With the exception of the YA3D-2Q which initially had a test boom in place of the refueling probe, all were fitted with the probe. The seven-man crew consisted of pilot, navigator, and DECM operator in the cockpit, and an ECM evaluator and three ECM operators in the fuselage compartment.

Over the years, ten A3D-2Qs (redesignated EA-3Bs after September, 1962) have been lost in accidents, one has been salvaged, and one is stored. (Recently, the cost of bringing BuNo 146449 back into service, has been estimated at $4 million, more than double the original price of the aircraft!) At the end of 1985, 13 others remained operational with VQ-1 and VQ-2 or were in the maintenance pipeline.

With no replacements in sight[1], the remaining EA-3Bs continue to fly much needed, but highly classified, ELINT and SIGINT missions from shore bases and during detachments aboard carriers of the Sixth and Seventh Fleets.

In 1961, the internal electronics of the A3D-2Q included AN/ALA-3 pulse analyzer, AN/ALR-8 countermeasures receiver, AN/APA-69 direction finder, AN/APA-74 signal analyzer, and three types of radar receivers (AN/APR-9, AN/APR-13 and AN/ARC-5). Cameras were provided at each operation station to record scope patterns for intelligence evaluation and another camera was used for photographing the ground. By retrofit prior to 1961, the aircraft had also received an AN/APX-175 radar set and three DECM systems (AN/ALQ-35, -41 and -51). Another retrofit, consisting in the installation of wing pylons as developed for the TA-3B to carry training stores, enabled EA-3Bs to be fitted with external ECM

---

[1] Beside the obvious parts problems associated with supporting single-aircraft detachments on carriers, EA-3B carrier operations will create standardization problems as, after the retirement of C-1As and reserve F-4s, this aircraft will be the last to use bridle launch gear.

BuNo 142673, the fifth A3D-2Q, was photographed on March 10, 1960, during service tests at NATC Patuxent River. Although only 25 A3D-2Q/EA-3Bs were built, this version has had the longest operational life of any "Whale". In early 1987, after the loss of BuNo 144850 in an unsuccessful barricade landing aboard the USS "Nimitz" (CVN-68) on January 25, 12 EA-3Bs were still operated by two Fleet squadrons, VQ-1 at NAS Agana, Guam, and VQ-2 at NAS Rota, Spain.

pods (such as the AN/ALQ-31).

Among the numerous upgrades since made to the EA-3B systems, those covered by Airframe Service Change ASC 284 (removal of the fin tip radome) and Airframe Change AFC 448 (installation of the ALR-40 *Seawing* system in an enlarged ventral canoe) deserve to be mentioned. The larger ventral *Seawing* canoe installed in accordance with AFC 448 reduced clearance with the deck during carrier operations and thus necessitated supplemental suitability evaluation prior to finalizing the production configuration. With BuNo 144849 outfitted with a prototype *Seawing* radome, tests were conducted at NATC Patuxent River, and aboard USS *America* (CVA-66) and USS *John F. Kennedy* (CVA-67) between June, 1968 and January, 1969.

In 1984, after having been upgraded several times since first fitted to service aircraft, the ALR-40 was composed of the following components/subsystems: ALR-40 antenna, receivers, controls, and indicators; ALR-63 IFM system (after AFC 474), ALR-44 countermeasures receiving system (after AFC 503), tuners, controls, and indicators; IP-1159A/A pulse analyzer indicator; APA-69 direction finder indicator; and APR-13 tuner. At that time, DECM systems fitted to the EA-3B included the ALQ-41, ALQ-51A, and ALQ-55.

Several unusual configurations are noteworthy. To photograph the re-entry of Soviet missiles in the Pacific in 1963-64, BuNo 146449 was to have received long focal length cameras in a glazed nose, but this modification was not realized. Instead, while operated by VQ-1 from Shemya AFB in 1964, with one Army officer and three enlisted men serving as equipment operators in the fuselage compartment, BuNo 146449 was fitted with a camera in place of the plane captain's seat. Originally, NADC Johnsville had designed a camera installation to be located in the starboard fuselage window but this installation did not prove satisfactory and was replaced with the camera shooting through the canopy. At a later time, BuNos 142673 and 146449 were fitted with phased array antennas on their ATM doors for use in Navy study of re-entry vehicles. Other EA-3Bs (again including BuNo 146449) were later fitted with a series of blade antennas atop the fuselage and LORAN-type "towel racks" on the fuselage sides.

The A3D-2T, the Trainer Version

In August, 1955, ComHatWing One requested a dual control A3D for checking out pilots during FIP as many of its first pilots were to be transferred from the patrol community and thus would lack jet experience. This request, however, was denied as the development of a pilot trainer version of the A3D-1 was not judged feasible due to time delays and engineering complications. Instead, HatWing One and Two's pilots were trained on Douglas F3D-2Ts and Lockheed TV-2s. The conversion of some A3D-1s to dual control configuration was later reported in the foreign press, but this rumor appears unfounded as old timers in the A-3 community cannot recollect dual control *Skywarriors* and Navy archives are void of any mention of such a conversion.

Although the Navy was confident that pilots could be trained without dual control *Skywarriors*, it soon recognized the need for a bombardier trainer version of the A3D.[2] Following issuance of a CNO confidential letter dated February 23, 1956, authorizing work on this version, a configuration conference was held at El Segundo on March 12/13 with COMAIRLANT, COMAIRPAC, BuAer, ComHat Wing One, BAR, and Douglas being represented. The agreed configuration provided for the use of an "A3D-2B" airframe (with CLE wings and refueling probe), a crew of nine (reduced to eight prior to finalization of the design), a master bombing station, seven (later five) repeater scopes, and the capability of dropping training bombs. Douglas was requested to submit a proposal including outline specifications, cost estimate, and delivery schedule, and to build a mock-up of the recommended configuration. Meanwhile, the Navy prepared an amendment to Contract NO(a)s 55-205, as substituting 12 bombardier trainer aircraft for an equal number of A3D-2Qs already on order would speed up deliveries and reduce costs.

In May, 1956, three weeks behind the originally scheduled date, the mock-up of the bombardier trainer was inspected and found highly satisfactory, thus enabling the Navy to include 12 A3D-2Ts (BuNos 144856 to 144867) in Contract NO(a)s 57-181 which, in the meantime, had been awarded in lieu of the NO(a)s 55-205 amendment. The first of these aircraft was airborne in May, 1959, and the last accepted in June, 1960. BIS trials for the A3D-2T were conducted at NATC Patuxent River between November, 1959 and March, 1960, clearing this version for initial service with VAH-3.

Redesignated TA-3Bs in September, 1962, the trainers (many of which had the "flat" nose and dove tail, without DECM systems, retrofitted over the years) and their VIP transport conversions (described anon) have seen much use. One of these aircraft, BuNo 144858, was fitted at one time with AN/ASB-12 radar and operated by RVAH-3 to train North American RA-5C *Vigilante* bombardier-navigators; later, this aircraft was modified back to the usual TA-3B standard.

One trainer and one transport conversion have been lost in accidents, a trainer has been salvaged, and the high-time transport conversion was used by Vought in 1982-83 for stress and fatigue tests to clear the remaining *Skywarriors* for service beyond their initially planned service life. The remaining aircraft were still operational at the end of 1985, with many having flown in excess of 10,000 hours, a remarkable achievement for an aircraft still being subjected to the harsh carrier environment. Five TA-3Bs were then operated by VAQ-33 to perform its mission as the A-3 Fleet Replacement Squadron, one was bailed to Grumman for use as an avionics testbed in support of

[2] Pending availability of the *Skywarrior's* bombardier-navigator trainer, at least one Lockheed P2V-3W *Neptune* (BuNo 124829) was fitted with an A3D-1 nose and radar system and used by VAH-3 for bombardier-navigator training. Other *Neptunes* without such an elaborate modification were also used in the early training program until a sufficient number of A3D-1s was available.

VQ-1 TA-3B, 144860, during a temporary stopover at Misawa AB, Japan. The TA-3B designator replaced the A3D-2T designator during September, 1962, when a standardized aircraft designating system went into effect for all military services. The VQ-1 tail art is noteworthy.

Photographed on December 3, 1959, after being modified as a staff transport by O&R Norfolk, BuNo 142672 was then still designated A3D-2Q even though most of its electronic equipment had been replaced by a plush interior. It is seen flying the flag of Vice Adm. Robert B. Pirie, then DCNO (Air).

Later in its life, BuNo 142672 became the only "Skywarrior" to carry the VA-3B designation identifying it as staff transport. It was lost at seat 125 miles NNW of Guam on January 23, 1985, while flown by the commanding officer of VQ-1.

BuNo 144857, a VIP-configured TA-3B, is shown undergoing overhaul at NARF Alameda on December 14, 1973. In late 1986, it was flown as a staff transport by COMFLELOGSUPPWING DET WASHINGTON DC, but was due to be replaced by a Gulfstream C-20B.

Although configured as a staff transport and bearing the markings of a transport squadron, VR-1, this TA-3B still retained its trainer designation so as not to offend money-conscious members of Congress! During 1986, this aircraft was refitted at NARF Alameda prior to assignment to VQ-1.

the F-14D development program, and two transport-configured aircraft were flown by CFLSW Det at NAF Washington.

For its original primary mission as a trainer for bombardiers and navigators, the A3D-2T had a normal crew of eight, consisting of pilot, assistant pilot, instructor and five trainees. The ASB instructor sat on the starboard side of the pressurized cabin and operated the master radar and bombing systems (with a blister fairing for the bombing periscope installed beneath the starboard fuselage side, instead of beneath the cockpit as on the bombers). A celestial navigation trainee sat in the cockpit, aft of the pilot but facing forward as opposed to the aft-facing third seat in other models, and four bombardier-navigator trainees sat in the fuselage compartment (three on the port side and one of the starboard side behind the instructor) and were provided with a navigation table and a repeater scope with related controls. Mission equipment included bomb assessment camera and recorder in a turret-like fairing (without provision for the guns) and two wing pylons (one outboard of each engine) for a 3,343-lb. (1,516 kg.) T-65 training shape or a 750-lb. (340 kg.) Aero 8A practice bomb dispenser carrying four MK 89 bombs. There was no provision for internal carriage of stores. Recognition features peculiar to this version are its periscope blister beneath the starboard side of the forward fuselage and its square fuselage windows (four on the port side and three on the starboard side).

The Transport Versions

In a report dated November 22, 1954, Douglas provided limited details on a design study for a military staff transport version of the *Skywarrior*. The proposed aircraft, with a pressurized fuselage compartment in an "A3D-2B" airframe, was to have been capable of carrying up to 14 military passengers in airline type seats on transcontinental routes such as from NAS North Island, California, to NAS Norfolk, Virginia. Two of the passenger seats were in the cockpit, aft of the pilot and assistant-pilot, with four rows of seats in the fuselage compartment (each row had two seats on the port side, an access aisle, and one seat on the starboard side). On this version, the lower entry and escape door was to be used only for emergency exit and a transport type door was provided on the port side. The Navy did not then have a requirement nor did it have the necessary funding for a jet transport but, 18 months later, the CNO expressed a definite interest in a transport version of the *Skywarrior* if such an aircraft could be obtained by modifying an existing airframe at minimal cost.

In compliance with the CNO's request, Douglas engineered a minimum modification of the A3D-2Q to obtain a transport. A mock-up of the proposed arrangement was inspected at the factory in January, 1957. This modification was such that the aircraft could be converted back to a -2Q configuration in a short time. To that effect, the transport aircraft was to retain the ventral canoe and fin tip radome of the electronic reconnaissance aircraft (both being later deleted; windows were later fitted on the port side as on the TA-3Bs). Less than one week after the mock-up had been inspected by the Navy, Pan American Airways' representatives were authorized to study it, as the airline was interested in acquiring a few of these aircraft to again experience with high-speed high-altitude jet aircraft before the entry into service of its Boeing 707s and Douglas DC-8s. Naught came from this proposed civilian use of the *Skywarrior* as Douglas could not deliver the aircraft required by Pan American before the 707s became available.

BuAer approved the A3D-2Q transport mock-up and the CNO authorized the conversion of the fourth A3D-2Q (BuNo 142672) as a staff transport for his use and that of senior DoD and Navy officials. However, to keep costs down and to avoid embarrassing arguments with Congress, the actual conversion work was assigned to O&R Norfolk. Readied in March, 1959, and accepted seven months later, this aircraft initially retained its A3D-2Q designation in spite of the fact that most of its electronic equipment had been replaced by a rather plush interior. Later on, its true role was admitted more freely and the aircraft was redesignated VA-3B (the V prefix indicating that it was a staff transport). No other transport conversions of the *Skywarrior* bore the VA-3B designation. Placed in SARDIP (Striken Aircraft Reclamation and Disposal Program) at Davis-Monthan AFB on April 9, 1974, the VA-3B was retrieved from MASDC on September 27, 1980, and sent to NARF Alameda to be returned to service with VQ-1 at NAS Agana, Guam. It crashed on January 23, 1985, while flown by the commanding officer of this unit.

After the usefulness of a jet-powered staff transport had been proven with this modified A3D-2Q, the Navy decided to obtain five additional aircraft by modifying A3D-2T airframes (BuNos 144857, 144860, 144863, 144864, and 144865) to the staff transport configuration. After September, 1962, to avoid congressional criticism, these aircraft retained the TA-3B designation of the unconverted trainers. Typically, these aircraft were fitted with three rows of airline type seats (two on the port side and one on the starboard side) in the fuselage compartment, a galley, and a lavatory; if required, the seats on the port side could be replaced by two bunks.

One of these aircraft crashed in 1974 while operated by VQ-2 and another was used by Vought for Navy-funded stress and fatigue testing after it had reached the end of its useful life. At the end of 1985, BuNos 144857 and 144865 were still operated by CFLSW-Det at NAF Washington to transport Navy and other government senior personnel as well as foreign dignitaries on official visits to the United States. Occasionally, these two aircraft were also used for humanitarian purposes to fly urgently needed medicine and medical equipment.

During the Southeast Asia War, the feasibility of obtaining another transport version of the *Skywarrior* was briefly considered by the Navy as it had a requirement for a fast COD transport to deliver fully assembled jet engines and other urgently needed parts to carriers operating in the Gulf of Tonkin. In this instance, the aircraft to be modified were to have been A-3Bs and the conversion was to affect only the bomb bay which was to receive an hydraulic hoist and special fixtures to carry fully assembled J79 turbojets complete with their afterburner. Improvements in the Navy supply pipeline eliminated the need for this fast COD transport before any A-3Bs were modified to this configuration. A less drastic cargo-carrying capability has since been provided for the KA-3Bs of VAK-208 and VAK-308 which can be fitted with a plywood container in the forward half of the original A-3B's bomb bay.

The Proposed Anti-ship and AEW Versions

Among the configuration studies undertaken in 1954 by Douglas were those for a version to be used against enemy ships and for an aircraft to fulfill the airborne early warning role. In the first instance, the manufacturer proposed modifying the A3D-2Q into a version for the detection, identification, tracking, and attack of enemy ships. The electronic equipment and interior arrangements were to be similar to those of the A3D-2Q, except that an AN/ASB-1A or more modern bombing radar system could replace the AN/ASB-1B search radar of the countermeasures aircraft. In addition,

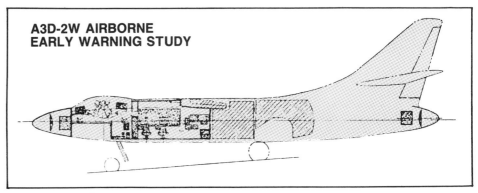

A3D-2W AIRBORNE EARLY WARNING STUDY

The first A3D-1, BuNo 130352, is shown as it appeared late in its life while operated at the Naval Missile Center Point Mugu. Unlike "Whales" used later by NMC/PMTC, this aircraft remained externally unmodified with the exception of the deletion of the fairing which had initially been fitted atop the fin.

Line-up showing eight of the "Strange Birds from Point Mugu". The aircraft in the foreground was modified during 1985 as a TNRA-3B for use in the NATOP training role by VQ-2. During 1986 PMTC still operated one NA-3B, one RA-3B, and three NRA-3Bs but it was rumored that some of these were to be transferred to FEWSG.

Probably the most distinctive and most photographed NMC/PMTC "Skywarrior", BuNo 144825 was fitted with its big nose by Grumman in 1960 when the aircraft was modified to test the pulse doppler radar and AAM-N-10 "Eagle" missile control system. It is seen here at NAS Willow Grove, Pennsylvania, in October, 1966.

BuNo 142667, built as the third A3D-2P, was still operated by the Pacific Missile Test Center at NAS Point Mugu in late 1986. In this view, the aircraft has been fitted with a seeker for use in a missile development program. Like many other PMTC "Whales", this NRA-3B has been fitted with a pylon beneath each wing.

This PMTC line up shows four NRA-3Bs and one NA-3B (side number 74, BuNo 138938, the second aircraft from the front). Noteworthy are the different nose shapes as each aircraft was modified to take part in a specific test and evaluation program. As a testbed, the "Skywarrior" has proved to be a most useful and reliable platform for which no replacement is yet available.

this version was to have carried an offensive armament comprised of two 2,500-lb. (1,134-kg.) "T" series stores, one beneath each wing.

Also based on the A3D-2Q airframe, the A3D-2W was a proposed airborne early warning aircraft with AN/APQ-50 search and fire control radar sets in the nose and tail cone. Its flight crew was to be located in the upper forward crew compartment and, in addition to the standard flight and navigation equipment, the pilot and assistant-pilot were each to be provided with scopes for the fire control unit mounted in the nose of the aircraft. Four *Sparrow* air-to-air missiles were to be carried beneath the wings and fired by the pilot. Two AEW operator stations were located in the lower aft section of the crew compartment and were provided with controls and scopes for the AN/APQ-50 and ECM equipment permitting earlier detection and location of radio emissions within the frequency range of 4,300 to 10,750 mc.

Neither one of these proposed versions found favor with BuAer and both remained on the drawing board.

The Strange Birds from Point Mugu and Other Odd Whales

Not surprisingly in view of the aircraft's good performance and capacious fuselage well adapted to carrying large electronic packages, the Navy Missile Center (NMC), later redesignated Pacific Missile Test Center (PMTC), at NAS Point Mugu evinced interest in acquiring *Skywarriors* at an early stage. Its wish was granted in 1959 when the replacement of A3D-1s by A3D-2s in deployable squadrons enabled the transfer of BuNo 135410 to Point Mugu. Since then, no fewer than 13 *Skywarriors* have been operated by NMC/PMTC and five of these aircraft were still used by PMTC at the end of 1985.

The following aircraft are known to have flown with NMC/PMTC:

Model	BuNo	Side No.	Remarks
NA-3A	130352	352	Lost in 1965.
NEA-3A	130363	Not known	Salvaged in January, 1970.
NA-3A	135409	72	Salvaged in May, 1975.
NA3D-1	135410	None	Lost in flight accident on January 22, 1959, when it collided with the water.
NA-3A	135411	Not known	May have been used only by Hughes, not by NMC. In storage at MASDC since January, 1975.
NA-3A	135418	70	Damaged in December, 1974. Now on display at the Naval Air Museum, Pensacola.
NA-3B	138938	74	Current.
KA-3B	138944	73	Now serving with VAQ-34.
NRA-3B	142667	71	Current.
NRA-3B	144825	75	Current.
RA-3B	144833	Not known	Current.
NRA-3B	144834	79	Now serving with VQ-2.
NRA-3B	144840	78	Current.
TA-3B	144867	77	Now bailed to Grumman.

For more than 25 years, NMC/PMTC has employed its A-3s for an amazing variety of experiments in support of development programs for air-launched missiles (e.g., *Phoenix* and *Tomahawk*), laser devices, and electronic warfare (ECM, ECCM, and ESM) systems, as well as for providing ECM threat simulation in support of training exercises of the Third Fleet. At first the Center's *Skywarriors* were externally unmodified but, as more complex weapons and systems were tested, they began sprouting an incredible array of pylons (either beneath the outer wing panels or on the lower sides of the forward fuselage), nose radomes and sensors, tail cone modifications, and external cameras.

Undoubtedly, the oddest of these *Skywarriors* is the still current Side Number 75 (BuNo 144825) with its "big nose." Modified in 1960 by Grumman for testing the pulse doppler radar and AAM-N-10 *Eagle* missile control system being developed by a Bendix/Grumman team for the D-970 *Missileer* (an earlier described derivative of the A3D) and the later Douglas F6D-1, BuNo 144825 was added to the Center's fleet in 1963. In addition to its large radome, this aircraft was at first characterized by the installation in the fuselage of a cooling system with external piping on the starboard side of the forward fuselage. Later on, it was fitted with a store pylon beneath the wings and had its tail cone repeatedly modified to house test radomes, recording instruments, and additional cooling devices as needed for the tests then being performed.

Test and Evaluation programs conducted by the Center with its A-3s are too numerous to be listed; moreover, many remain classified. However, the accompanying photographs provide an overview of many of the no longer classified activities.

The other principal operator of odd-looking A-3s has been Hughes Aircraft Company. Its first *Skywarrior* was BuNo 135427, an NA-3A, which was modified in 1964-65 as a flying testbed for the *Phoenix* AMCS (Airborne Missile Control System) being developed for the Grumman F-111B. Modifications included installation of an F-111B radome to house the *Phoenix* AMCS antenna, radar, and infrared subsystems. To accommodate these and associated instrumentation, a 40-inch (1.02-m) extension of the aircraft's nose was necessary and this, along with the radome unit, increased overall length by 56 inches (1.42 m). Two 8-inch (0.20-m) "I" beams installed crosswise in the bomb bay protruded on either side to form support pylons for missiles used in pit ejection and airborne launch tests. A 20 KVA gas turbine power unit was added to supply the required power for the added instrumentation and systems. These included a test instrumentation package with a 150-channel frequency modulation-pulse code tape recording system located in the bomb bay, a photo-recording system with six externally-mounted launch cameras, a time/data correlation system, a telemetry receiving system, and miscellaneous support equipment.

The first full scale *Phoenix* test employing all functions of the missile control system was made on September 8, 1966. Installed in BuNo 135427, the AMCS located a jet target drone and, having locked on it at long range, launched the *Phoenix* which scored a probable kill. Participating in additional tests for the next four years, this NA-3A was salvaged in August, 1970.

Another NA-3A (BuNo 135411) was briefly used by Hughes but was placed in storage at MASDC in January, 1975. Lacking the pressurized fuselage compartment of the versions, the A-3As were not ideally suited as testbeds. Accordingly, Hughes briefly operated an NRA-3B (BuNo 142667, which afterward became Side Number 71 with PMTC) prior to obtaining a TA-3B in October, 1968. This long-lived aircraft (BuNo 144867 which was also operated by PMTC as Side Number 77)

The NRA-3B, BuNo 142667, was briefly used by Hughes Aircraft to test the guidance system for the "Phoenix" missile. Later this aircraft was transferred to PMTC and, assigned the side number 71, it was still in use at NAS Point Mugu during 1986.

BuNo 135411, a NA-3A which was bailed to Hughes for use in radar development programs, was fitted at one time with a distinctive "thimble" nose radome. While operated by Hughes, this testbed had a high visibility scheme with red forward fuselage, wing tips and vertical tail surfaces. Replaced by the NRA-3B illustrated on the preceding page, this aircraft has been kept in storage at Davis-Monthan AFB since January, 1975.

first served with Hughes to flight test the AN/AWG-9 radar for the Grumman F-14A. Often upgraded, notably with the installation of an APU and liquid cooling system in the rear fuselage and tail cone, BuNo 144867 normally had a five-man test crew (pilot, crew chief, instrumentation engineer, and two flight test engineers). In 1985, the aircraft was scheduled to be modified once again for use by Grumman in testing the Hughes AN/APG-71 digital signal processing radar for the F-14D version of the *Tomcat*.

The ERA-3B, the Electronic Aggressor Version

Following the decommissioning of VAP-61 in July, 1971, a number of its RA-3Bs were taken over by VQ-1 while other aircraft became available for conversion as electronic aggressors for use by VAQ-33. Bearing the ERA-3B designation, the modified aircraft (initial modification being performed under AFC 460) were no longer capable of operating from carriers as their empty weight exceeded the maximum carrier landing weight but retained the hook for emergency field arrestment. They were fitted with an Aero 7A external store pylon and ejector rack beneath each wing and Electronic Warfare (EW) systems grouped in three categories: Electronic Countermeasures (ECM), Electronics Support Measures (ESM), and Defensive Electronic Countermeasures (DECM). To provide power for all this additional electronic equipment, these aircraft were also fitted with four external ram air turbines (RATs), two on each side of the forward fuselage. Early ERA-3Bs used RATs similar to those on the AN/ALQ-76 pods but later aircraft received RAT's from AN/ALQ-99 pods.

From the onset, the ERA-3B had a five-man crew with pilot, navigator, and plane captain in the cockpit and two ECM evaluators in the pressurized fuselage cabin (with an additional upper escape hatch). Its ECM compartment housed part of the equipment and had two swivel seats for the ECM operators. Other equipment was fitted in the original photo-flash bomb bay of the RA-3Bs, a ventral canoe, a fin tip housing, and an extended tail cone of revised geometry and housing chaff dispensers (initially AN/ALE-2 and later AN/ALE-43). External wing racks were wired for AN/ALQ-76 pods.

Over the years, the ERA-3B system have been upgraded several times, a process which is anticipated to continue. Hence, the following list of equipment is given only as representative of aircraft used by VAQ-33 in the early to mid-eighties.

Electronic Warfare Systems: WJ-8570 microwave direction finding and omni-directional antenna system; AN/ALR-43(V) countermeasures receiving system; AN/ULA-2 pulse analyzer system; AN/ALT-32H, AN/ALT-27V, and AN/ALQ-76 pod-mounted system electronic countermeasure transmitting systems; AN/AQM-37A missile target launching system; AN/ALQ-31A electronic countermeasures pod; AN/ALE-2 chaff dispenser system; AN/ALE-25 countermeasures equipment dispensing set; AN/ALE-29 countermeasures chaff dispensing set; and AN/DLQ-3B multiple ECM threat environment systems.

Navigation, Automatic Flight Control, and Communication Systems: AN/ASN-50 attitude and heading reference system; AN/APN-153 radar navigation system; AN/ASN-41 navigation computer system; AN/APN-70 LORAN navigation system; AN/ARA-50 direction finding

ERA-3B, BuNo 144832, of VAQ-33 and a TA-4J, BuNo 158518, of VC-1 north of Oahu, Hawaii, during an exercise in May, 1978. Noteworthy are the two RATs (Ram Air Turbines) on the fuselage sides and the AN/ALQ-76 pods beneath the wings. Additional ECM equipment is fitted in the photo-flash bomb bay and in the ventral canoe. The extended tail cone houses AN/ALE-2 chaff dispensers. The tail code, GD, identifies the Fleet Electronic Warfare Support Group (FEWSG) and is used by both VAQ-33 and VAQ-34.

system; AN/ARC-51A dual UHF communication system; LTN-211 Omega navigation system; as well as other systems common with those fitted to standard RA-3Bs.

For several years, only four ERA-3Bs (BuNos 144827, 144832, 146446, and 146447), operated by VAQ-33, were in existence. Then, in anticipation of the establishemnt of VAQ-34 in March, 1983, four additional RA-3Bs (BuNos 142668, 144838, 144841, and 144846) were retrieved from storage at MASDC and overhauled at NARF Alameda. At that time, these ERA-3Bs were fitted with a new 19-ft (5.79-m) ventral radome with a honeycomb core, which was developed by Boeing Military Airplane Co. to house the AN/ALT-40. The new radome has also been retrofitted to earlier ERA-3Bs. The active and passive electronic warfare equipment for the new ERA-3Bs was supplied by EM Systems, Hewlett-Packard Co., Lundy Electronics & Systems, Inc., Raytheon Co., Scientific Communications, Inc., and Watkins-Johnson Co. Payload integration for the later ERA-3Bs was done in Waco, Texas, by Electro-Space Systems under contract from the Naval Electronic Systems Command. In 1985, two more RA-3Bs (BuNos 142666 and 142669) were being brought to ERA-3B standard to enable VAQ-33 and VAQ-34 to have four aircraft while two other ERA-3Bs are being overhauled.

Skywarrior Production Summary
The Versions

	Contract NOa(s)	FSN	BuNos	No. of aircraft	Remarks
YA3D-1P		9259	130358	(1)	(1)
YA3D-1Q		9257 and 9261/9264	130356 and 130360/130363	(5)	(2)
A3D-2P	55-205	11582 and 11729/11732	142256 and 142666/142669	5	(3)
	57-181	12071/12093	144825/144847	23	(4)
	57-181	12398/12399	146446/146447	2	(5)
A3D-2Q	55-205	11583 and 11733/11736	142257 and 142670/142673	5	(3)
	57-181	12094/12101	144848/144855	8	(4)
	57-181	12400/12411	146448/146459	12	(5)
A3D-2T	57-181	12102/12113	144856/144867	12	(5)

Remarks:
(1) Modified from 7th A3D-1 airframe.
(2) Modified from 5th, and 9th through 12th A3D-1 airframes.
(3) All but first A3D-2P and A3D-2Q (BuNos 1442256 and 142257) delivered with refueling probe.
(4) Fitted with refueling probe during production.
(5) With CLE wings and refueling probe.

A Douglas TA-4J of VC-1 formates with ERA-3B, BuNo 144827 of VAQ-33 during a 1978 RIMPAC operation. The ERA-3B is carrying only one AN/ALQ-76 ECM pod, visible under its port wing.

ERA-3B, BuNo 146447, of VAQ-33, during a temporary stopover at NAS Oceana. This aircraft was equipped with a conventional RAT-powered AN/ALQ-76 ECM pod under its port wing and a chaff dispensing unit in its extended tail cone.

ERA-3B, BuNo 144841, of VAQ-34, provides a good illustration of the aircraft's wing and vertical tail surfaces in their foided position. The vertical fin and rudder assembly fold to starboard, and the wing outer panel sections fold inboard toward the aircraft centerline. Folding is accomplished through movement imparted by hydraulic actuators. Visible in this photo is the tailcone ejection slot through which cut chaff is ejected into the aircraft slipstream.

Chapter 6: Operational Service

Nine A3D-1s are seen outside of the El Segundo Division prior to completion and delivery to the Navy. After delays were incurred prior to and during flight trials, the A3D program rapidly gained tempo following initial service use. Four heavy attack squadrons, VAH-1 and VAH-3 with the Atlantic Fleet and VAH-2 and VAH-4 with the Pacific Fleet, received their "Skywarriors" during 1956. The type's first operational cruise was made from January 15 until July 22, 1957 by VAH-1 aboard the USS "Forrestal" (CVA-59).

Following the start of BIS trials in May, 1955, planning for the *Skywarrior's* service debut gained impetus and reached fruition while A3D-1s and A3D-1Qs participated in the Fleet Indoctrination Program. Basically, the plans formulated at that time called for the temporary assignment of A3D-1s to Heavy Attack Squadrons, commencing with VAH-1 in LANT Fleet and VAH-2 in PAC Fleet. As soon as sufficient A3D-2s became available to equip additional squadrons and replace the earlier aircraft in the first squadrons, the A3D-1s would be transferred to the RAG or used for sundry experimental purposes. Likewise, in service with VQ-1 (PAC) and VQ-2 (LANT), A3D-2Qs were to replace the interim A3D-1Qs as soon as possible. In the photographic reconnaissance role, the sole YA3D-1P was to undergo operational evaluation with VAP-62 (LANT), but A3D-2Ps were to be the primary mission aircraft in this squadron and its sister unit in PAC Fleet, VAP-61.

These plans were implemented beginning on March 31, 1956, when VAH-1 obtained its first five A3D-1s. Already, Douglas and North American Aviation, Inc. were hard at work on competing designs to provide a supersonic successor to the *Skywarrior* (the Navy issuing a Letter of Intent for two North American XA3J-1s in September, 1956). By existing standards, the A3D was expected to have a service life of 10 to 12 years. Certainly, no one was then farsighted or foolish enough to dare predicting that three decades later (including a nine-year war) *Skywarriors* would still equip Fleet squadrons. Yet, 30 years after reaching the Fleet and more than six years after its intended successor was phased out, more than 50 *Skywarriors* are still going strong!

Early Service Life: Nuclear Deterrence on the High Seas

Upon entering service, *Skywarriors* were assigned to squadrons[1] of Heavy Attack Wing One for operations with the Atlantic Fleet and Heavy Attack Wing Two for service with the Pacific Fleet. The honor of introducing the A3D-1 into service went to Heavy Attack Wing One, with five aircraft being ferried from NAS Patuxent River to NAS Jacksonville on March 31, 1956, by VAH-1 crews led by their commanding officer, Cdr. Paul F. Stephens. During the following month, two A3D-1s went to Evaluation Squadron Five (VX-5) and a first aircraft was delivered to VAH-2 on the 30th to initiate *Skywarrior's* operations in PAC Fleet.

The *Skywarrior's* importance in contemporary Navy planning was stated as follows in a letter from Vice Adm. Thomas S. Combs, DCNO (Air), to Vice Adm. William L. Rees, Commander Air Force, U.S. Atlantic Fleet, on June 6, 1956:

> As you know the introduction of the A3D aircraft into the fleet is one of our most important jobs. I understand that a normal training period for the first VAH squadron would preclude deployment until the Fall of 1957. However, it is my understanding that a large percentage of the pilots in the first operational unit have had previous experience. Also, the A3D had a highly successful FIP at Patuxent River. In view of these factors it appears that some acceleration may be possible at this time.
>
> I have completed most of my testimony for the Symington Subcommittee. In this presentation I have cited the complements of the *Forrestal* and subsequent CVAs as including 12 A3Ds. The converted *Midway* class was also noted as capable of handling 9 A3Ds.
>
> In view of the great importance of exploiting this new capability at an early date, I would appreciate your views and in particular advice as to what assistance you may require in order to deploy A3Ds in *Forrestal* and *F.D.R.*

Prodded by the DCNO (Air) and drawing on their experience during FIP, VAH-1 crews soon reported excellent results from first operations in spite of a shortage of support items. Less than six months later, after all mandatory carrier suitability fixes had been incorporated in their A3D-1s, VAH-1 crews conducted carrier qualifications aboard the USS *Forrestal* (CVA-59) in preparation

This A3D-1 from VAH-4 was photographed just after leaving the starboard catapult of the USS "Shangri-La" (CVA-38) during carquals on November 22, 1959. Note the bridle falling below the aircraft just ahead of the bow. Another A3D-1 is being readied on the port catapult and two others are on the aft deck.

During the fall of 1956, when tension rose over the Soviet invasion of Hungary and the Anglo-French operations against Egypt, VAH-1 was split into two detachments, one going aboard USS "Forrestal" (CVA-59) and one aboard USS "Saratoga" (CVA-60). BuNo 135420 is seen recovering aboard CVA-60 on November 15, 1956.

On March 31, 1956, the first five A3D-1s assigned to VAH-1 were ferried from NAS Patuxent River to NAS Jacksonville. One of these aircraft, BuNo 135421, is seen here during this historic flight. The tail code and modex have not yet been applied and the number 21 on the nose is the last two digits of the BuNo.

Bearing the tail code of CVG-17, BuNo 135441, an aircraft from VAH-3, is seen here about to hook a wire during carquals aboard USS "Saratoga" (CVA-60) sometime in 1957. VAH-3 was the second LANT squadron to receive "Skywarriors".

This tanker-configured A3D-2 (BuNo 138932) of VAH-1 was photographed during a JATO takeoff at the Lambert-St. Louis IAP. Note that on this occasion only six bottles, three on each side, were used. The tail code AG identifies CVG-7 with which VAH-1 made three deployments aboard USS "Independence" (CVA-62).

A-3B, BuNo 142633, of VAH-2, from the USS "Coral Sea", during a temporary stopover at NAS Alameda, California, on May 13, 1966. The open bomb bay provides access to the aircraft's inflight refueling system. Discernible on the fuselage side to the rear of the canopy is the unit patch.

for their first deployment then scheduled for January, 1957, as part of Air Group One (CVG-1). Once again, the squadron claimed encouraging results, with Cdr. Stevens commenting that:

> The first operations were extremely successful for the first time aboard with a new aircraft, particularly considering that this was a heavier aircraft, with all the attendant complexities.

Indeed, the apparent ease with which the *Skywarrior* was introduced into carrier operations was a fortunate development as the Navy was forced to take precautionary measures when world tensions rose sharply following both the Anglo-French operations against Egypt, beginning on October 31, 1956, and the Soviet invasion of Hungary three days later. At the time, VAH-1 was divided into two detachments, one going aboard *Forrestal* for a one-month cruise (November 7-December 12) in the eastern Atlantic and the other being kept ready for deployment aboard the USS *Saratoga* (CVA-60). Fortunately, tension soon subsided and VAH-1 returned to NAS Jacksonville to complete preparations for its first scheduled deployment to the Mediterranean aboard CVA-59. This cruise, which sent without major incident and lasted from January 15 until July 22, 1957, marked the start of the *Skywarrior's* service life in its intended role: providing deterrence while deployed aboard carriers.

Being the first A3D operational squadron and playing a role in the late 1956 contingency operations, VAH-1 had rightfully attracted more attention than the three other *Skywarrior* squadrons which came into being during that year. VAH-3, with LANT, received its first A3D-1 on October 8, 1956, and first deployed aboard the USS *Franklin D. Roosevelt* (CVA-42) on July 9, 1957. With PAC, VAH-2 obtained its first A3D-1 on April 30, 1956[2] and its detachment Bravo first deployed aboard the USS *Bon Homme Richard* (CVA-31) on July 11, 1957. The corresponding dates for VAH-4, also with PAC Fleet, were October, 1956, and February, 1958 (Det Delta aboard the USS *Hancock* [CVA-19]). Other LANT Heavy Attack Squadrons received their first *Skywarriors* in January, 1957 (VAH-9); July, 1957 (VAH-5); November, 1957 (VAH-11); March, 1958 (VAH-7); and February, 1961 (VAH-13). In PAC, VAH-8 was established in May, 1957, and received A3D-2s beginning the following August, VAH-6 converted from AJ-2s to A3D-2s in early 1958, VAH-123 had *Skywarriors* when it was established in June, 1959, and VAH-10, the last Heavy Attack Squadron to be formed, got its first A3D-2s in June, 1961.

Notwithstanding sanguine reports on the introduction of the A3D to carrier operations, most squadrons ran into serious problems during carquals and initial deployments. At that time, most senior crews in the A3D community had little jet/CVA training as they were former VC or VP pilots used to flying propeller-driven aircraft from shore bases. Consequently, boarding rate was appallingly low and many CAGs and carrier COs

---

[1] In a strangely naive attempt at disinformation, the VAH squadrons were originally classified as "land-based mining squadrons." During the same period of time, the Air Force listed its high-altitude Lockheed reconnaissance aircraft as U-2s in an equally vain effort to have them pass as "utility aircraft." In view of the prowess of Soviet intelligence, it is doubtful that either of these deceptions fooled the Russians for very long. Yet, to this day, we often tend to resort to similar "white lies" to cover the obvious. With or without detente, we should not underestimate so lightly the ability of the Soviets to see through such covers and turn them around to make us appear devious in the eyes of the unsuspecting. There is a good reason for labeling "intelligence" the gathering of information on potential enemies!

[2] On April 4, 1966, Edgar D. Mitchell (the first VAH-2 pilot to complete night carquals in May, 1957) was selected to join the astronaut team. He had been preceded by another *Skywarrior* pilot, Roger B. Chaffee, formerly with VAP-62, who had been selected on October 18, 1963.

were dismayed. Fortunately, at the behest of Rear Adm. J. D. Ramage, more pilots with jet experience were added and the Heavy Attack Squadrons emphasized tactical and carrier training, thus eliminating the early training shortcomings.

As A3Ds operated more and more frequently aboard carriers, the type gained a nickname that almost universally replaced its official name. While, in theory, as many as 27 *Skywarriors* could be accommodated on the flight and hangar decks of CVA-19 class angled-deck carriers, there was no doubt that the size of the aircraft caused much concern to CAGs. In those days, few if any of these senior flight officers had prior experience with heavy attack aircraft and, consequently, most CAGs bitterly complained that A3Ds occupied "too much valuable real estate" and began referring to these aircraft as "whales." Far from retaining a derisive connotation, the nickname stuck and to this day is used with much affection by most.

After entering service, the A3D-1s of VAH-1 began sporting the tail code TB (the initials of Capt. J. T. "Tommy" Blackburn, the commanding officer of Heavy Attack Wing One) while those of VAH-2 were coded BF. These non-standard codes soon gave place to a system whereby, when the squadrons did not apply the code of the Carrier Air Group to which they were assigned, those belonging to Heavy Attack Wing One used codes starting with the letter G and those of Heavy Attack Wing Two, codes beginning with the letter Z. These codes were GH (VAH-1), GJ (VAH-3), GK (VAH-5), GL (VAH-7), GM (VAH-9), GN (VAH-11), and GP (VAH-13) in Heavy Attack Wing One, and ZA (VAH-2), ZB (VAH-4), ZC (VAH-6), ZD (VAH-8), and ZE (VAH-10) in Heavy Attack Wing Two. The code NJ was used by VAH-123 as it was part of Replacement Carrier Air Group Twelve.

Skywarrior squadrons initially took care of their own pilot and bombardier-navigator training requirements and, for that purpose, operated a number of Douglas F3D-2Ts, Lockheed P2V-3Bs and TV-2s, and Grumman F9F-8Ts alongside their A3D-1s and A3D-2s. However, as more squadrons were formed or converted from North American AJ-2s, the Fleet recognized the need for specialized conversion units and thus set up HATULANT and HATUPAC (Heavy Attack Training Units Atlantic/Pacific) in June, 1957. HATULANT was absorbed into VAH-3 in June, 1958, whereas HATUPAC was redesignated VAH-123 on June 29, 1959; on these dates, VAH-3 and VAH-123 respectively became the Replacement Training Squadron (RAG) for the Atlantic Fleet and the Pacific Fleet. To fulfill this role, the two squadrons added A3D-2Ts beginning at the end of 1959. The responsibility for training *Skywarrior* crews became solely that of VAH-123 in January, 1964, when VAH-3 graduated its last A-3 class to concentrate on training crews for the North American A-5A and RA-5C. Responsibility for the A-3 crew training was successively passed on to VAQ-130 in February, 1971, when VAH-123 was decommissioned, to the Naval Air Reserve Unit (NARU) Alameda in June, 1974, when VAQ-130 began transitioning to the Grumman EA-6B, and finally to VAQ-33 in October, 1977. With these various units, the training syllabus evolved as the role of the *Skywarrior* was changing. By 1961 emphasis had shifted from high altitude operations to low-level navigation and bombing, and four years later refueling procedures and techniques became major elements of the training program.

Due in part to the care with which the first crews were trained, the *Skywarrior* had a relatively safe period of initial operations. The first service accident occurred on February 9, 1957, near Mayport, Florida, when both engines of BuNo 135457 flamed out due to fuel starvation. The aircraft came down intact in shallow water but its pilot,

Heavy Attack Squadron Three "Sea Dragons" made a single cruise with "Skywarriors" prior to being merged with the Atlantic Fleet's Heavy Attack Training Unit (HATULANT) in June, 1958, to become the A3D Replacement Training Squadron for the Atlantic Fleet. The VAH-3 aircraft illustrated here is an A3D-1 (BuNo 135419).

Beginning at the end of 1959, VAH-3 supplemented its motley inventory of training aircraft with A3D-2Ts. BuNo 144857 is seen taxiing with the cable of its drag chute still taut and an Aero 8A practice bomb dispenser beneath its right wing. The first tail code letter identifies Heavy Attack Wing One.

In conjunction with a note in an A3D-1 Standard Aircraft Characteristics Chart stating that "A total of 27 airplanes can be accommodated aboard a CVA-19 class carrier," this view of 25 "Whales" on the VAH-3 ramp at NAS Sanford would have been enough to give recurring nightmares to any self-respecting CAG!

Prior to A3D-2Ts becoming available, various models of the Lockheed "Neptune" were used to train A3D bombardier-navigators. This P2V-3W, BuNo 124289, assigned to VAH-3 is unusual in being fitted with the nose radome of the A3D-1 and practice bomb racks beneath the wing.

Taken in 1959 during VAH-6's first cruise aboard the USS "Ranger" (CVA-61) as part of CVG-14, this photograph shows an A3D-2 leaving the carrier's angled deck. A second aircraft is about to be catapulted. The tail code ZC identifies the third squadron of Heavy Attack Wing Two, not the Carrier Air Group.

In September, 1959, VAH-7 made the first operational A3D to A3D refueling, with BuNo 142664, a tanker-configured A3D-2, transferring fuel to BuNo 142665. The tanker pilot was Cdr. "Speed" Moreland and that of the receiver was LCdr. Tom Stewart.

BuNo 142237, an A-3B of VAH-8 "Fireballers", photographed trapping aboard USS "Midway" (CVA-41) in 1964. Between August, 1958, and December, 1967, VAH-8 made six cruises aboard CVA-41, one of which was a war cruise, and two war cruises aboard the USS "Constellation" (CVA-64).

To show off their flying prowess to the fighter community, four A-3B crews are seen making a formation fly-by next to the USS "Midway" (CVA-41). After making three war cruises, the "Fireballers" of VAH-8 were disestablished at NAS Whidbey Island on January 17, 1968.

The tail code ZC carried by this A3D-2 (BuNo 142640) of VAH-6 helps date this photograph to the first half of 1959 when this squadron deployed aboard the USS "Ranger" (CVA-61) as part of Carrier Air Group Fourteen for a WestPac cruise.

LCdr. Pollock, was killed when his parachute failed to open. Unfortunately, safety soon deteriorated; seven aircraft were lost in operational accidents during 1957 and eight went down in 1958. Moreover, the next three years saw a worsening of the situation as operational losses reached a total of 14 aircraft in 1959, 11 in 1960 and 10 in 1961. Training and maintenance procedures were then tightened and losses were reduced to seven aircraft in 1962, went back to 10 in 1963 and finally dropped to four in 1964.

In the early years, after initial training and carrier qualifications gave place to regular deployments (in squadron strength aboard the large carriers and in four-aircraft detachments aboard the small CVAs) with the Sixth and Seventh Fleets, the Heavy Attack Squadrons contributed significantly to the Nation's deterrence when several crises (e.g., U.S. intervention in Lebanon in July, 1958, Sino-Communist activities in the Taiwan Straits in August, 1958, and the Cuban Missile Crisis in October, 1962) threatened to lead to major wars. Deployable squadrons were then flying A3D-2s and their primary mission was high altitude attack against strategic targets, carrying either one 6,650-lb (3,016-kg), three 2,035-lb (923-kg) or three 1,300-lb (590-kg) special stores[3].

Most combat training sorties were planned to duplicate eventual wartime missions, with single aircraft being assigned individual targets. At first these sorties were flown on internal fuel only, but after September, 1959, following the entry into service with VAH-7 of the first A3D-2s fitted with refueling probe and tanker package, training to reach more distant targets was begun. In addition, "buddy bombing" tactics were developed by VAH-6. In this instance, A3D-2 bombers were assigned strategic targets while light attack aircraft, refueled in flight by A3D-2 tankers and carrying tactical nuclear weapons, were to help the bombers' penetration by striking heavy defensive points along the bombers' path. Another new tactic practiced in the early sixties by the Heavy Attack Squadrons called for low-level penetration and "loft" bombing to offset improvements in enemy defenses. (The effectiveness of Soviet anti-aircraft missiles against high-flying aircraft had been proved on May 1, 1960, when the Lockheed U-2 of Francis Gary Powers was shot down near Sverdlovsk.)

Although deterrence was the A3D-2's raison d' être during the late fifties and early sixties, fortunately for the sanity of the crews, preparing for a possible nuclear holocaust had not been their only activity. In fact, in pre-Southeast Asia War years, these crews made several notable flights including the following:

September 3, 1956: Two A3D-1s of VAH-1 launched from the USS *Shangri-La* (CVA-38) off the coast of Oregon and landed in Oklahoma City; they had flown 1,543.3 miles (2,483 km) in two hours 32 minutes and 39.7 seconds.

June 6, 1957: Two A3D-2s of VAH-9 were launched from the USS *Bon Homme Richard* (CVA-31) off the coast of California and landed aboard the USS *Saratoga* (CVA-60) off the east coast of Florida. Upon landing, the crews were greeted by President Eisenhower.

July 16, 1957: Two A3D-2s of VAH-2 flew from NAS Moffett Field, California, to NAS Barbers Point, Hawaii, in four hours and 45 minutes.

October 11, 1957: The California to Hawaii record was bettered by a VAH-4 A3D-2 with a con-

---
[3] Although the A3D-1 had been designed to carry a single special store of large dimensions and weight, an ECP for multiple carriage of special weapons had been submitted by Douglas in December, 1956. Approval of this engineering change enabled the A3D-2 to carry several of the smaller and lighter nuclear devices developed in the late fifties and early sixties. Douglas designed and built the casing and control systems for several of these weapons, with Sandia Corporation providing the explosive charge and nuclear elements.

trol tower to control tower flight from San Francisco to Honolulu in four hours 29 minutes and 55 seconds.

October 27-November 1, 1959: To demonstrate the feasibility of a North Pacific track for flights to and from the Orient, an A3D-2 of VAH-4 used the Great Circle route. The aircraft flew the 4,218 miles (6,787 km) from NAS Whidbey Island, Washington, to NAS Atsugi, Japan, in 11 hours and 29 minutes, including a 45-minute refueling stop at Adak in the Aleutians. This track saved nearly 2,800 miles (4,505 km) over the United States to Japan route via the Hawaiian Island and other refueling points in the Pacific previously used by naval aircraft. On the return, the aircraft departed from Misawa AFB, Japan, and proceeded to Whidbey with a refueling stop at King Salmon AFB, Alaska; flight time was eight hours and 24 minutes.

March 13-15, 1960: Nine A3D-2s of VAH-8 were launched from the USS *Midway* (CVA-41) while she was sailing some 400 miles off Japan and proceeded to NAS Whidbey Island, Washington, via NAS Barbers Point, Hawaii, and NAS Alameda, California. Total distance was 4,800 naut. miles (8,890 km) and flight time 10.9 hours.

May 24, 1961: As part of the celebration of the 50th anniversary of U.S. Naval Aviation, the Navy organized Project *LANA* (L for the roman numeral 50, and ANA for Anniversary of Naval Aviation). For *LANA*, five F4H-1s competed in the Bendix Trophy race in an attempt to break the transcontinental speed record of 3 hours and 8 minutes set in November, 1957, by an Air Force McDonnell RF-101C. The *Phantoms*, which took off from Ontario, California, and landed at NAS Brooklyn, New York, were refueled by A3D-2s from VAH-4 near Albuquerque, New Mexico, from VAH-1 over St. Louis, Missouri, and from VAH-9 over Pittsburgh, Pennsylvania. One of the F4H-1s was forced out of the race after encountering problems over Albuquerque and another one straggled far behind due to refueling difficulties. The winning F4H-1 crew (Lt. Richard F. Gordon and Lt(jg) Bobbie R. Young) flew the 2,445.9 miles (3,935.5 km) in two hours 45 minutes; the times for the other two crews were two hours and 57 minutes, and three hours and three minutes[4].

August 23-24, 1963: As part of Project *Stormfury*, a joint Weather Bureau-Navy experiment to determine whether the energy patterns of large storms could be changed, an A-3B seeded hurricane *Beulah* with silver iodine particles.

December 1, 1964: Three A-3Bs of VAH-2 were catapulted at night from the USS *Coral Sea* (CVA-43) while she was moored to the pier at Ford Island, Hawaii.

The strategic deterrent role of the A3D began waning when the USS *George Washington* (SSBN-598), the first submarine to be armed with a full load of 16 *Polaris A-1* fleet ballistic missiles, became operational on December 15, 1960, and departed Charleston, North Carolina, on its first cruise. Furthermore, the A3D's usefulness in the nuclear strike role was threatened by the impending entry into service of the North American A3J-1 *Vigilante*. Nevertheless, two more Heavy Attack Squadrons were established in 1961 (VAH-13 at NAS Sanford, Florida, on January 3, and VAH-10 at NAS Whidbey Island, Washington, on May 1) to bring the A3D attack force to a peak of 13 squadron, including 11 deployable HATRONs and two RAG squadrons. Thereafter, the number of Heavy Attack Squadrons decreased steadily: VAH-7 converted to A3J-1s in August, 1961, and to RA-5Cs in late 1964; VAH-1 acquired its first

The USS "Midway" (CVA-41) passed under the Golden Gate Bridge on August 16, 1958 at the start of a seven-month WestPac cruise. On the deck can be seen A3D-2s of VAH-8 along with other CVG-2 aircraft: F8U-1s of VF-211, F3H-2s of VF-64, FJ-4Bs of VA-63, AD-6s of VA-65, AD-5Ws of VAW-11, and AD-5Ns of VA(AW)-35.

As the Aero 21B tail turret proved to be unreliable, the 20mm M3L cannon were frequently removed in service as shown in this view of BuNo 138911, an A3D-2 of VAH-13. The "Bats" were commissioned at NAS Sanford on January 3, 1961, and made only two cruises prior to converting to the RA-5C in 1964.

A "Project LANA" (L = 50th, ANA = Anniversary of Naval Aviation) A-3B, BuNo 138929, of VAH-9 while assigned to the USS "Saratoga" (CVA-60). "Project LANA" aircraft refueled five F4H-1s for the 1961 Bendix Trophy race. Markings are typical for VAH-9 aircraft with the unit patch visible on the forward fuselage.

---

[4] The F4H-1 record was broken by an Air Force Convair B-58 on March 5, 1962, with a time of two hours and 56.8 seconds. However, the B-58 had entered the trap at altitude and speed whereas the LANA Phantoms had been timed from brake release.

Taken on March 28, 1961, this clean view of an A3D-2P (BuNo 144843) of VAP-62 illustrates the photo-reconnaissance version of the "Skywarrior" in the configuration in which it was delivered (i.e.; pointed nose radome, tail turret and rear fuselage fittings for JATO bottles). What appears to be a porthole on the fuselage side just below the canopy's rear window is the squadron insignia. Bearing U.S. Army markings, BuNo 144843 was still in use at the White Sands Missile Range in the mid-eighties.

A-5As in September, 1962, and was redesignated RVAH-1 in September, 1964; VAH-3 graduated its last *Skywarrior* class in January, 1964; and VAH-5, VAH-6, VAH-9, VAH-11, and VAH-13 became Reconnaissance Heavy Attack Squadrons between May, 1964, and July, 1966.

On September 18, 1962, shortly after this series of conversions had been initiated, the *Skywarriors* in service with HATRONs were redesignated A-3As, A-3Bs, and TA-3Bs. With deployable squadrons, the normal complement of which was then a mix of A-3Bs with standard wings and A-3Bs with CLE wings and tanker package, training was progressively switched from nuclear attack and mining operations with conventional weapons as the Kennedy Administration placed greater emphasis on preparedness for fighting localized conflicts. In addition, these squadrons were increasingly used to provide pathfinding and tanking for other carrier-based aircraft. As it turned out, the revised training proved of great value when the United States was drawn into the Vietnam conflict.

Although VAP-62 had received the sole A3D-1P for a brief period of operational evaluation commencing October 14, 1957, the *Skywarrior* was not regularly used by reconnaissance squadrons until A3D-2Ps began replacing North American AJ-2Ps in VCP-61 at NAS Agana, Guam, in September, 1959, and in VAP-62 at NAS Jacksonville, Florida, one month later. Redesignated VAP-61 after its Vought F8U-1Ps were transferred to VFP-63 in July, 1961, the PAC unit flew its A3D-2Ps (RA-3Bs after September, 1962) almost exclusively from shore bases on routine missions prior to the start of the Southeast Asia War; its aircraft only made occasional short-term WestPac deployments aboard carriers of the Seventh Fleet. On the other hand, during the period preceding and immediately following the initial American involvement in Vietnam, LANT's HATPHOTORON was given numerous unusual assignments including the following:

February 19, 1962: Cruising at 40,000 ft. (12,190 m), LCdr. William Warde and his VAP-62 crew photographed John H. Glenn, Jr.'s *Friendship 7* Atlas-Mercury rocket and capsule from a distance of barely one mile. Thereafter, VAP-62 was called to photograph other manned and unmanned launches from Cape Canaveral/Cape Kennedy.

October-November, 1962: During the Cuban Missile Crisis, VAP-62 flew a variety of missions including photography of Soviet vessels transporting military cargo to and from Cuba. In recognition of its work during this crisis, the squadron was cited in April, 1963, by the Secretary of the Navy for "outstanding and meritorious service in performance of classified aerial photographic reconnaissance of paramount military importance to the security of the United States."

November 11, 1963: This date marked the first RA-3B non-stop transatlantic flight, from NAS Jacksonville to NS Rota, in 7.3 hours with only one inflight refueling.

October 1964: For the U.S. Army Map Service, RA-3Bs did photographic mapping of all the shore line on the United States side of the Great Lakes.

November 1964: A two aircraft detachment operating from Howard AFB in the Panama Canal Zone flew high and low altitude mapping missions of different lakes, streams, and coast lines, and obtained initial cartographic photography of Costa Rica, Honduras, Nicaragua, and Panama for the Inter-American Geodetic Survey. This work was again performed one year later.

July 28-October 1965: A detachment operating from San Juan, Puerto Rico, took part in a new phase of Project *Stormfury*, the experimental seeding of tropical storms over the Western Atlantic and Eastern Caribbean.

Spring of 1966: Special infrared photography and comparative water penetration analysis was undertaken as part of Project *Autec*.

June 1966: The squadron obtained complete photographic coverage of all shipyard installations in Boston, Charleston, Norfolk, Philadelphia and Portsmouth.

For less than two years, A3D-2Ps were also assigned to Composite Reconnaissance Squadron Sixty-Three (VCP-63) at NAS Miramar, California. However, on July 1, 1961, the squadron lost its *Skywarrior* and, equipped solely with Vought F8U-1Ps, was redesignated VFP-63.

Another achievement of the photographic reconnaissance version of the *Skywarrior*, undertaken by a Douglas test crew prior to the entry into service of the A3D-2P, deserves to be mentioned. This March, 1959, mission provided horizon-to-horizon photo-mapping coverage of a 450-mile (270-km) strip from Wendover AFB, Utah, to Edwards AFB, California, along the proposed flight path of the North American X-15 research aircraft. The data was subsequently used by X-15 pilots to identify reference points and to familiarize themselves with emergency landing areas.

The A3D-1Q version entered service with VQ-1 at MCAF Iwakuni, Japan, on November 3, 1956, and one month later joined VQ-2 in Port Lyautey, Morocco. With these two squadrons, the "Queer Whales" supplemented elderly Martin P4M-1Qs as a TASES (Tactical Airborne Signal Exploitation System) component to provide ELINT and SIGINT respectively for PAC and LANT. Compared to the old *Mercators*, the A3D-1Qs were substantially more capable not only due to their more sophisticated equipment and higher performance but also because of their ability to operate from carriers when areas of interest were unreachable from hospitable foreign land bases. The delivery of A3D-2Qs beginning in late 1959 enabled VQ-1 and VQ-2 to phase out their P4M-1Qs and interim A3D-1Qs (but not until three of the five -1Qs had been lost in operational accidents). Redesignated EA-3Bs in September, 1962, and with their systems upgraded as technology and funding allowed, the definitive electronic reconnaissance version of the *Skywarrior* has seen continuous service for over 26 years.

Operation Wetwing: Skywarriors in Vietnam

In the early sixties, after President Kennedy had sent U.S. advisors to South Vietnam and authorized a plan for covert operations against North Vietnam to counter increased Communist subversive activities in the Indo-Chinese peninsula, Seventh Fleet carriers on WestPac deployments made more and longer cruises in the South China Sea. By 1964, tension had reached near fever pitch and the inevitable happened: on August 2, 1964, the USS *Maddox* (DD-731) was attacked by Soviet-built P-4 class torpedo boats. In what became known as the Tonkin Gulf Incident, the USS *Ticonderoga* (CVA-14), joined two days later by the USS *Constellation* (CVA-64), made the first retaliatory strikes against North Vietnam. Among the aircraft used in this initial action, although only in a standby tanking role, were A-3Bs of VAH-4 Det Bravo aboard CVA-14 and VAH-10 aboard CVA-64. From then on, *Skywarriors* were part of the American war effort until the cessation of hostilities on August 15, 1973.

Although for most of the war A-3Bs (and later KA-3Bs and EKA-3Bs) were primarily used by HATRONs and TACELRONs for aerial refueling, at first they also flew bombing and mining sorties over both North and South Vietnam. (See Chapter 8 for an account of the first bombing mission over the North.) Even when fitted with the refueling package, and early on not all A-3Bs deployed to the Tonkin Gulf were so fitted, *Skywarriors* could still carry a hefty punch; they were the only Navy aircraft capable of internally carrying 2,000-lb/ 908-kg bombs[5]. Moreover, some A-3Bs, such as those of VAH-4 Det Golf aboard the USS *Oriskany* (CVA-34) in April-December 1965, were fitted with the A-4 *Skyhawk* gunsights for dive bombing attacks. However, a shortage of bombs in the 1965-67 period, the A-3's vulnerability to SAMs and MiGs, and their greater usefulness in the tanking role soon led to their exclusive use as tankers.

Regular use of A-3B tankers enabled Air Wings to mount missions against distant targets, such as in Laos, or to approach closer targets using other than the most direct routes in order to avoid concentrated defenses. It was, however, the "unscheduled" use of the tankers which finally won the heart of CAGs. As fighting intensified, more and more fighter and attack aircraft were returning to their carriers short on fuel or with damaged tanks to be saved only by timely "top-ups" from Whales.

Although the A-3Bs were supplemented by A-4s and A-7s fitted for buddy refueling, *Skywarriors* were the best suited for the tanking task until the advent of the Grumman KA-6D. Accordingly, as detailed in Chapter IV, NARF Alameda undertook to convert *Skywarriors* into KA-3B dedicated tankers and EKA-3B combination tanker/tactical jammer aircraft. KA-3Bs and EKA-3Bs first deployed with specially established TACELRONS in the fall of 1967 and, within one year, had fully replaced A-3Bs. HATRONs operated KA-3Bs from 1968 until August, 1970 when VAH-10 Det 14 returned to Whidbey Island at the end of the last deployment by a Heavy Attack Squadron. As the war went on, KA-3Bs and EKA-3Bs were replaced by KA-6Ds and EA-6Bs aboard large carriers, but they continued to operate from small carriers until the fighting ended in 1973.

Whatever the version, *Skywarrior* tankers proved extremely valuable in saving human lives and valuable aircraft[6]. A few examples set forth their usefulness:

January 1966: To supplement carrier-based aircraft, VAH-4 began maintaining a shore-based Detachment Yankee for operations from NAS Cubi Point and DaNang AB.

May 31, 1967: During operations over the North, two A-3Bs of VAH-4 found themselves critically short of fuel and proceeded to an orbiting Air Force KC-135A (which had already refueled two Air Force F-104Cs) to tank up. Along with the A-3Bs came two F-4Bs and two F-8Es which were equally thirsty. In fact, the F-8Es could not wait until the A-3Bs had been refueled and hooked up to the *Skywarriors* as they were taking on fuel from the KC-135A. The triple level refueling proved timely and all aircraft were able to return to their carriers or bases. Altogether, the KC-135A had transferred 49,900 lb. (28,190 liters) of fuel and, with the help of the two A-3Bs, had saved eight aircraft in one sortie[7].

August 13-14, 1967: During a 36-hour search and rescue operation, VAH-8 transferred a record 234,000 lb. (132,200 liters) of fuel (an amount equivalent to the full fuel load of a McDonnell Douglas DC-10-30)!

During the Southeast Asia War, VAH-4 lost one A-3B to MiGs and one to unknown combat related causes. Thirteen A-3B/KA-3B/EKA-3Bs were also lost in operational accidents. Nevertheless, these losses were considerably exceeded by the number of aircraft saved by tanking Whales.

Whereas for VAH and VAQ *Skywarriors* carrier operations were the rule and shore detachments[8] the exception, the reverse was true for the EA-3Bs and RA-3Bs of VQ and VAP squadrons. Thus, while VQ-1, VQ-2, VAP-61, and VAP-62 occasionally sent *Skywarriors* to operate from carriers on Yankee Station, they flew most of their combat missions from DaNang AB in the Republic of Vietnam and NAS Cubi Point in the Philippines. Supplemented by VQ-2 Detachment Bravo, which maintained two EA-3Bs and crews at DaNang AB from 1965 until 1969, VQ-1 conducted electronic reconnaissance missions throughout the war with its EA-3Bs. In addition, until carrier-based tactical aircraft were fitted with self-contained radar and missile jamming systems (e.g., ALQ-51, ALQ-81, ALQ-100, APR-25 and APR-27) in early 1967, EA-3Bs operating from carriers in the Gulf of Tonkin flew deep into North Vietnam to provide warning and jamming for the strike aircraft. Following the disestablishment of VAP-61 on July 1, 1971, VQ-1 inherited the responsibility for photographic and weather reconnaissance missions with RA-3Bs. Usually operating above the effective ceiling of conventional AAA and seldom having to venture in SAM or MiG country after 1966, the EA-3Bs ended the war without incurring a single combat loss and with only four aircraft lost in operational accidents (none of which occuring in the war zone).

When America was drawn into the Vietnamese conflict, an urgent need for target and mapping photography suddenly developed. Although the RA-3B lacked the performance required for target reconnaissance in heavily defended areas, it had several advantageous features not found in faster tactical reconnaissance aircraft (the Air Force RF-101A/C and then RF-4C, the Navy RF-8A/G and RA-5C, and the Marine RF-8A and then RF-4B). Notably, the reconnaissance version of the *Skywarrior* could carry a greater variety of cameras including large ones particularly well suited for night photography and cartographic survey and an ample supply of flash cartridges and flash bombs. Furthermore, infrared films were soon used by the RA-3Bs for camouflage detection, thus adding to the aircraft's usefulness.

Making use of these capabilites, VAP-61, which was supplemented by three RA-3Bs and four crews from VAP-62 Detachment Pacific, provided most of the cartographic survey needed in support of the war effort. Its main mission, however, was night reconnaissance along the North Vietnamese infiltration network, the infamous Ho Chi Minh Trail. This task, which necessitated flying at relatively low altitude (as low as 1,000 ft/305 m when using the electronic flash over jungle terrain) over well-defined tracks, proved a dangerous one, and VAP-61 lost four aircraft to enemy action (two other RA-3Bs were lost in operational accidents during the war). In an attempt to minimize the RA-3B's visual detectability, several VAP-61 aircraft were camouflaged; some (e.g., BuNos 144840 and 144847) were repainted glossy black overall while others (e.g., BuNos 144831, 144834, and 144846) had a wrap-around camouflage of FSN 36118 sea gray, FSN 36251 medium gray, and FSN 36440 light gull gray. The camouflage proved quite effective but did not prevent BuNo 144847 from failing to return from a mission over

---

[5] Typical conventional bomb loads for the A-3B in bomber and tanker configurations were as follows:

	Bomber	Tanker
500-lb bomb (high drag)	12	8
1,000-lb bomb (high drag)	6	3
2,000-lb bomb (high drag)	4	2
500-lb bomb (low drag)	6	4
1,000-lb bomb (low drag)	4	3
2,000-lb bomb (low drag)	3	0
374-lb mine	12	N/A
500-lb mine	12	8
1,000-lb mine	6	3
2,000-lb mine	4	2

[6] The often hectic needs of fighter and attack aicraft during Alpha strikes involving several tankers and a larger number of other aircraft led to the application of special markings to identify quickly individual tanker aircraft. For example, VAH-4 Det 31 aboard the USS *Bon Homme Richard* (CVA-31) painted its aircraft thusly: one had a diagonal red stripe on top of its wings, around its rear fuselage, and atop the fin; the second had two blue stripes in the same positions; and the third had no special markings. While operating from the USS *Kitty Hawk* (CVA-63), VAQ-131 used diagonal black stripes on the rear of the fuselage of its KA-3Bs and EKA-3Bs. One VAQ-131 aircraft had a single stripe, another had two stripes, and so on; the largest number of stripes, six, was seen on NH 616, BuNo 142632.

[7] For his role in this spectacular feat, SAC was preparing to court martial the KC-135A pilot for breaking the SAC rule forbidding three-level refueling. Timely arrival of a Navy "well-done" letter saved the KC-135A pilot from being court martialed; instead, Major Casteel was decorated by the AF!

An RA-3B of VAP-61 is seen landing aboard the USS "Kitty Hawk" (CVA-63) on February 25, 1964 when the carrier was making its second WestPac cruise. Heavy Photographic Reconnaisse Squadron Sixty-one received its first A3D-2P during November, 1959, and flew photo-reconnaissance "Skywarriors" until decommissioned on July 1, 1971.

Photographed in the snow at Misawa AB, Japan, on New Year's Day 1975, this TA-3B (BuNo 144860) of VQ-1 was far from the warmer clime of the unit's base at NAS Agana, Guam. The PR tail code assigned to VQ-1 has lead to the nickname "Peter Rabbit" being used unofficially by the "World Watchers."

Route Package 2 on January 1, 1968. This, however, was the last *Skywarrior* combat loss.

With the decommissioning of VAP-62 in October, 1969, and that of VAP-61 in July, 1971, the reconnaissance mission was taken over by VQ-1. Along with the squadron's EA-3Bs, the RA-3Bs then remained in use until after the American withdrawal from the war.

Air operations over South and North Vietnam ended on January 27, 1973, and those over Laos on April 17. Missions over Cambodia continued until August 15, 1973, when, in accordance with a congressional edict, all U.S. combat operations in Southeast Asia were ended. At that time, the only *Skywarriors* still active in the area were the EA-3Bs and RA-3Bs of VQ-1 and two EKA-3B detachments of VAQ-135, respectively embarked on the USS *Coral Sea* (CVA-43) and USS *Hancock* (CVA-19). After the end of fighting in Southeast Asia, VAQ-130 sent KA-3B/EKA-3B detachments to the Western Pacific aboard the USS *Oriskany* (CVA-34) and USS *Ranger* (CVA-61). The latter's return to homeport on October 10, 1974, with the EKA-3Bs of VAQ-130 Det Four, ended regular *Skywarrior* deployments aboard carriers. Since then, only the EA-3Bs of VQ-1 and VQ-2 have operated aboard carriers of the Sixth and Seventh Fleets.

Skywarriors Do not Die...and Do not Fade Away

In the early 1970s, when RA-3Bs were about to be retired and KA-3Bs and EKA-3Bs were being replaced aboard carriers by Grumman KA-6Ds and EA-6Bs, the days of the *Skywarrior* appeared to be numbered. While, indeed, many of these aicraft were then stricken or placed in storage at MASDC, a number of KA-3B/EKA-3Bs found a new lease on life when assigned to two newly created Reserve squadrons (VAQ-208 and VAQ-308) or used in sundry test and support roles and several RA-3Bs were then modified into ERA-3Bs and NRA-3Bs for use respectively by the Fleet Electronic Warfare Support Group (FEWSG) and PMTC.

Commissioned in 1970, the Alameda-based Reserve squadrons had been planned to have a dual role: tanking and electronic countermeasures. However, both units soon concentrated on providing refueling and pathfinding not only as part of their assigned Air Wing (CVWR-20 and CVWR-30) operations but also in support of deployments by Navy and Marine active units.[9] In recognition of this specialized role, VAQ-208 and VAQ-308 were redesignated Aerial Refueling Squadrons (VAK-208 and VAK-308) in 1979. In early 1986, the KA-3Bs of these two squadrons remained the only Navy aircraft providing support for transpacific and transatlantic aircraft movements. This situation, however, may change in the near future if the Navy obtains congressional approval to acquire used commercial aircraft and modify them into tankers.

Since the early 1970s, the EA-3Bs, of VQ-1 and VQ-2 (for which no replacement has yet been identified) have continued to fulfill a stealthy and vital role as their effectiveness has been increased several times as the result of avionic and electronic upgrades. In addition to routine operations from shore bases, these aircraft are still deployed regularly aboard carriers operating in the Mediterranean Sea and the Atlantic, Indian, and Pacific Oceans. In September, 1985, EA-3Bs from VQ-2 monitored communications in the eastern Mediterranean area and thus played a vital, but discreet, role int he successful interception of the Egyptair Boeing 737 carrying Palestinian terrorists out of Egypt. Likewise, they were active prior to, during and after the April, 1986, confrontation with Libya.

In the fleet electronic warfare support role, the ERA-3Bs have similarly refused to die and, far from fading away, have seen their number increased to enable the formation of a second squadron (VAQ-34) in March, 1983.

Three decades after VAH-1 received its first A3D-1s, *Skywarriors* are still operated by two Fleet Air Reconnaissance Squadrons (VQ-1 at NAS Agana, Guam, and VQ-2 at NS Rota, Spain), two Tactical Electronic Warfare Squadrons (VAQ-33 at NAS Key West, Florida, and VAQ-34 at NAS Point Mugu, California), and two Aerial Refueling Squadrons (VAK-208 and VAK-308 at NAS Alameda, California). Other *Skywarriors* are operated by the Pacific Missile Test Center at NAS Point Mugu, California, the Naval Weapons Center at NAS China Lake, California, and a detachment of the Fleet Logistics Support Wing at NAF Washington, Andrews AFB, Maryland. Finally three aircraft are bailed to Grumman, Raytheon and Westinghouse.

If funding can be obtained, the indestructible *Skywarrior* may yet have a new life as a research aircraft. Seeking to develop a high-speed, Upper Surface Blowing (USB) research aircraft to extend into higher speed and altitude the work done with its QSRA (Quiet Short-haul Research Aircraft, a Boeing-modified de Havilland C-8A), the NASA Ames Research Center has evaluated the feasibility of modifying various aircraft. In 1983, Ames selected the A-3 as the most suitable vehicle. The *Skywarrior* has the necessary performance envelope and strong airframe, is compatible with the proposed use of four General Electric TF34 or Garrett ALF-502 turbofans (mounted above the wings to provide for upper surface blowing of the inboard *Coanda* flaps), and its high wing and fuselage-mounted landing gear would make conversion relatively easy. If approved, the conversion would retain the fuselage, undercarriage and vertical tail surfaces of the A-3 but would use a new wing with reduced sweep (15° to 25°), advanced airfoils and winglets, modified T-tail stabilator, and canard surfaces.

Altogether, at the end of 1985, 51 *Skywarriors* (13 KA-3Bs, two NA-3Bs, 13 EA-3Bs, two RA-3Bs, eight ERA-3Bs, four NRA-3Bs, one TNRA-3B, and eight TA-3Bs) were still acive. In addition, two

This EA-3B of Fleet Air Reconnaissance Squadron One is about to be launched from the USS "Kitty Hawk" (CV-63) on April 7, 1975, during Operation "Rimpac 75", a joint Canada, New Zealand, Australia, and United States exercise. Note that this EA-3B bears more discrete tail markings than does the TA-3B illustrated above.

---

[8] One big advantage for working "off the beach" (out of DaNang) was that tankers could operate at full weight—78,000 lb. versus 73,000 lb. max. catapult weight. Some dets made a "normal practice" of launching off the CVAs, tanking all aircraft in sight, and going into DaNang for turnaround. Two or three hours later, they launched at full weight and returned to the ship with 5,000 lb. more fuel to pass.

[9] On August 1, 1975, when returning from one of these pathfinding/tanking missions, a KA-3B of VAQ-208 flew the longest non-stop flight by a carrier-based aircraft. It covered 6,100 miles (9,815 km) from NAS Rota, Spain, to NAS Alameda, California, in just about 13 hours.

VAQ/VAK-208 and -308 have refueled all types of Navy aircraft equipped with a refueling boom. Less well-known is the use in the spring of 1973 of a VAQ-308 aircraft (BuNo 138929) for dry hook-ups with a Lockheed NT-33A (Air Force serial 51-4120) modified by Calspan Corporation to test "fly-by-wire" controls. Flown by G. Warren Hall, the NT-33A was fitted with a probe taken from a North American F-100 and bolted on its starboard side.

KA-3Bs (BuNos not known) were kept on ready reserve at MASDC and two RA-3Bs (BuNos 142666 and 142669) had been taken out of storage to be refurbished as ERA-3Bs.

Indeed, it is quite remarkable that almost 19% of all A-3s built were still in service (or being restored for service use) 30 years after the type's debut with the Fleet.[10] To have achieved a similar longevity record, over 1,100 Douglas SBD *Dauntlesses* would have had to be operational in 1970 and over 2,500 Vought F4U *Corsairs* would have had to be in military service at the end of 1972! Obviously, the rather flat progress curve in airframe and powerplant technology since the 1950s has had much to do with the *Skywarrior's* long life, preventing its early obsolescence. However, even so, the A-3 would have had a much shorter life had it not been for the ease with which it was adapted to fulfill new roles and its remarkably strong construction. Former SBD, A-1, and A-4 crews would certainly be ready to join A-3 crews in acknowledging that the El Segundo team surely knew how to build tough airplanes!

[10] The Air Force B-52 has set an even more impressive record as about 35% of the *Stratofortresses* built by Boeing were still in service (in their originally intended role) 30 years after the 93rd Bombardment Wing received its first B-52B. To achieve this spectacular feat, however, B-52s have had to go through several major rework programs to extend their structural life. No such costly program has been necessary to keep A-3s flying and making arrested landings.

This artist's rendering shows the proposed Upper Surface Blowing research aircraft as it would appear in flight over the distinctive dirigible hangar at NAS Moffett Field. NASA Ames Research Center still hopes to obtain the funding to modify an A-3 airframe and fit it with four turbofans to provide for upper surface blowing.

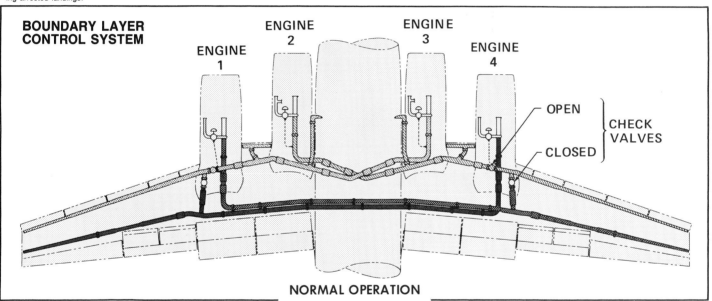

Retaining the fuselage, undercarriage, and vertical tail surfaces of the A-3, the USB research aircraft is to be fitted with T-tail stabilator, canard surfaces, and a wing of reduced sweep using advanced airfoils and winglets. Four Garrett ALF-502 or GE TF34 turbofans would replace the J57 turbojets powering standard A-3s and would provide for upper surface blowing of the inboard "Coanda" flaps.

SKYWARRIOR HATRON & TACELRON DEPLOYMENTS
LANT
1957-1973

Squadron/Air Wing	Aircraft	Carrier	Tail Code & Modex	Deployment Dates
VAH-1/CVG-1	A3D-1	USS *Forrestal* (CVA-59)	TB x	01/15/57–07/22/57
VAH-1/CVG-7	A3D-2	USS *Independence* (CVA-62)	AG 6xx	08/04/60–03/03/61
VAH-1/CVG-7	A3D-2	USS *Independence* (CVA-62)	AG 6xx	07/14/61–12/18/61
VAH-1/CVG-7	A3D-2	USS *Independence* (CVA-62)	AG 6xx	04/19/62–08/27/62
VAH-3/CVG-17	A3D-1	USS *F. D. Roosevelt* (CVA-42)	AL 3xx	07/09/57–03/04/58
VAH-5/CVG-10	A3D-2	USS *Forrestal* (CVA-59)	AK 5xx	09/02/58–03/12/59
VAH-5/CVG-8	A3D-2	USS *Forrestal* (CVA-59)	AJ 6xx	01/28/60–08/31/60
VAH-5/CVG-8	A3D-2	USS *Forrestal* (CVA-59)	AJ 6xx	02/09/61–08/25/61
VAH-5/CVG-8	A-3B	USS *Forrestal* (CVA-59)	AJ 6xx	08/03/62–03/02/63
VAH-6/CVW-8	A-3B	USS *Forrestal* (CVA-59)	AJ 6xx	07/10/64–03/13/65
VAH-9/CVG-3	A3D-2	USS *Saratoga* (CVA-60)	AC 1xx	2/ 1/58–10/ 1/58
VAH-9/CVG-3	A3D-2	USS *Saratoga* (CVA-60)	AC 5xx	8/14/59– 2/26/60
VAH-9/CVG-3	A3D-2	USS *Saratoga* (CVA-60)	AC 5xx	8/22/60– 2/26/61
VAH-9/CVG-3	A3D-2	USS *Saratoga* (CVA-60)	AC 5xx	11/28/61– 5/11/62
VAH-9/CVG-3	A-3B	USS *Saratoga* (CVA-60)	AC 5xx	3/29/63–10/25/63
VAH-10/CVW-1	A-3B	USS *F. D. Roosevelt* (CVA-42)	AB 61x	6/28/65–12/17/65
VAH-10 Det 66/CVW-6	A-3B	USS *America* (CVA-66)	AE 02x	1/10/67– 9/20/67
VAH-10 Det 42/CVW-1	KA-3B	USS *F. D. Roosevelt* (CVA-42)	AB	8/24/67– 5/19/68
VAH-10 Det 60/CVW-3	KA/EKA-3B	USS *Saratoga* (CVA-60)	AC 8xx	5/ 2/67–12/ 6/67
VAH-10 Det 62/CVW-7	KA-3B	USS *Independence* (CVA-62)	AG	4/26/68– 1/27/69
VAH-10 Det 59/CVW-17	KA-3B	USS *Forrestal* (CVA-59)	AA 61x	7/22/68– 4/29/69
VAH-10 Det 67/CVW-1	KA/EKA-3B	USS *J. F. Kennedy* (CVA-67)	AB 65x	4/ 5/69–12/21/69
VAH-10 Det 60/CVW-3	KA/EKA-3B	USS *Saratoga* (CVA-60)	AC 61x	7/ 9/69– 1/22/70
VAH-10 Det 59/CVW-17	KA/EKA-3B	USS *Forrestal* (CVA-59)	AA 61x	12/ 2/69– 7/ 8/70
VAH-10 Det 62/CVW-7	KA/EKA-3B	USS *Independence* (CVA-62)	AG	6/23/70– 2/ 1/71

(During this cruise, the detachment was redesignated VAQ-129 Det 62 on 9/ 1/70.)

Squadron/Air Wing	Aircraft	Carrier	Tail Code & Modex	Deployment Dates
VAH-11/CVG-1	A3D-2	USS *F. D. Roosevelt* (CVA-42)	AB	2/13/59– 9/ 1/59
VAH-11/CVG-1	A3D-2	USS *F. D. Roosevelt* (CVA-42)	AB 6xx	1/28/60– 8/24/60
VAH-11/CVG-1	A3D-2	USS *F. D. Roosevelt* (CVA-42)	AB 6xx	2/15/61– 8/28/61
VAH-11/CVG-1	A-3B	USS *F. D. Roosevelt* (CVA-42)	AB 6xx	9/14/62– 4/22/63
VAH-11/CVW-1	A-3B	USS *F. D. Roosevelt* (CVA-42)	AB 6xx	4/28/64–12/22/64
VAH-11 Det 8/CVW-1	A-3B	USS *Independence* (CVA-62)	AB 6xx	9/ 8/64– 5/11/65
VAH-11/CVW-8	A-3B	USS *Forrestal* (CVA-59)	AJ 6xx	8/24/65– 4/ 7/66
VAQ-129 Det 62/CVW-7	EKA-3B	USS *Independence* (CVA-62)	AG	6/23/70– 2/ 1/71

(This detachment had started the cruise as VAH-10 Det 62 but was redesignated VAQ-129 Det 62 on 9/ 1/70.)

Squadron/Air Wing	Aircraft	Carrier	Tail Code & Modex	Deployment Dates
VAQ-130 Det 42/CVW-6	EKA-3B	USS *F. D. Roosevelt* (CVA-42)	AE 01x	1/ 2/70– 7/27/70
VAQ-130 Det 4/CVW-3	EKA-3B	USS *Saratoga* (CVA-60)	AC 61x	6/17/70–11/11/70
VAQ-130 Det /CVW-6	EKA-3B	USS *F. D. Roosevelt* (CVA-42)	AE 80x	2/15/72–12/11/72
VAQ-130 Det 1/CVW-1	KA/EKA-3B	USS *J. F. Kennedy* (CVA-67)	AB 61x	4/16/73–12/ 1/73
VAQ-131/CVW-1	EKA-3B	USS *J. F. Kennedy* (CVA-67)	AB 61x	9/14/70– 3/ 1/71
VAQ-135 Det 1/CVW-6	EKA-3B	USS *F. D. Roosevelt* (CVA-42)	AE 8xx	2/12/71– 7/28/71
VAQ-135 Det 2/CVW-8	EKA-3B	USS *America* (CVA-66)	AJ 61x	7/ 6/71–12/16/71
VAQ-135 Det 4/CVW-1	EKA-3B	USS *J. F. Kennedy* (CVA-67)	AB 61x	11/18/71–10/ 5/72
VAQ-135 Det 2/CVW-17	EKA-3B	USS *Forrestal* (CVA-59)	AA 61x	9/22/72– 7/ 6/73

Assigned to CVW-6, VAQ-130 Det 42 deployed to the Med aboard USS "Franklin D. Roosevelt" (CVA-42) between January and June 1970. One of its EKA-3Bs is seen here over the Caribbean Sea in August, 1969, during a pre-deployment exercise.

A3B, BuNo 138918, of VAH-11 then stationed aboard the USS "Forrestal" (CVA-59). Gray and white markings are typical for type as are unit patch and tail code locations. Slight upward flex to wing under load is noteworthy.

SKYWARRIOR HATRON & TACELRON DEPLOYMENTS
PAC
PRE-VIETNAM WAR PERIOD: 1957-1964
POST-VIETNAM WAR PERIOD: 1973-1974

Squadron/Air Wing	Aircraft	Carrier	Tail Code & Modex	Deployment Dates
VAH-2 Det B/CVG-5	A3D-2	USS *Bon Homme Richard* (CVA-31)	ZA xx	7/10/57-12/11/57
VAH-2 Det M/CVG-9	A3D-2	USS *Ticonderoga* (CVA-14)	ZA xx	9/16/57-4/25/58
VAH-2 Det H/ATG-1	A3D-2	USS *Ticonderoga* (CVA-14)	ZA xx	10/4/58-2/17/59
VAH-2 Det E/CVG-19	A3D-2	USS *Bon Homme Richard* (CVA-31)	ZA x	11/1/58-6/18/59
VAH-2/CVG-15	A3D-2	USS *Coral Sea* (CVA-43)	NL 9xx	9/19/60-5/27/61
VAH-2/CVG-15	A3D-2	USS *Coral Sea* (CVA-43)	NL 6xx	12/12/61-7/17/62
VAH-2/CVG-15	A-3B	USS *Coral Sea* (CVA-43)	NL 6xx	4/3/63-11/13/63
VAH-4 Det D/CVG-15	A3D-2	USS *Hancock* (CVA-19)	ZB xx	2/16/58-10/3/58
VAH-4 Det C/CVG-11	A3D-2	USS *Shangri-La* (CVA-38)	ZB xx	3/8/58-11/22/58
VAH-4 Det C/CVG-11	A3D-2	USS *Shangri-La* (CVA-38)	ZB xx	3/9/59-10/3/59
VAH-4 Det L/CVG-21	A3D-2	USS *Lexington* (CVA-16)	ZB xx	4/26/59-12/1/59
VAH-4 Det D/CVG-15	A3D-2	USS *Hancock* (CVA-19)	ZB xx	8/1/59-1/18/60
VAH-4 Det E/CVG-19	A3D-2	USS *Bon Homme Richard* (CVA-31)	ZB xx	11/21/59-5/14/60
VAH-4 Det N/CVG-5	A3D-2	USS *Ticonderoga* (CVA-14)	ZB xx	3/5/60-10/11/60
VAH-4 Det F/CVG-14	A3D-2	USS *Oriskany* (CVA-34)	ZB xx	5/14/60-12/15/60
VAH-4 Det C/CVG-11	A3D-2	USS *Hancock* (CVA-19)	ZB xx	7/16/60-3/18/61
VAH-4 Det L/CVG-21	A3D-2	USS *Lexington* (CVA-16)	ZB xx	10/29/60-6/5/61
VAH-4 Det E/CVG-19	A3D-2	USS *Bon Homme Richard* (CVA-31)	ZB xx	4/26/61-12/13/61
VAH-4 Det B/CVG-5	A3D-2	USS *Ticonderoga* (CVA-14)	ZB xx	5/10/61-1/15/62
VAH-4 Det F/CVG-14	A3D-2	USS *Lexington* (CVA-16)	ZB xx	11/9/61-5/12/62
VAH-4 Det L/CVG-21	A3D-2	USS *Hancock* (CVA-19)	ZB xx	2/2/62-8/20/62
VAH-4 Det G/CVG-16	A3D-2	USS *Oriskany* (CVA-34)	ZB xx	6/7/62-12/17/62
VAH-4 Det E/CVG-19	A3D-2	USS *Bon Homme Richard* (CVA-31)	ZB xx	7/12/62-2/11/63
VAH-4 Det B/CVG-5	A-3B	USS *Ticonderoga* (CVA-14)	ZB xx	1/3/63-7/15/63
VAH-4 Det L/CVG-21	A-3B	USS *Hancock* (CVA-19)	ZB xx	6/7/63-12/16/63
VAH-4 Det G/CVW-16	A-3B	USS *Oriskany* (CVA-34)	ZB xx	8/1/63-3/10/64
VAH-4 Det E/CVW-19	A-3B	USS *Bon Homme Richard* (CVA-31)	ZB xx	1/28/64-11/21/64
VAH-6/CVG-14	A3D-2	USS *Ranger* (CVA-61)	ZC xx	1/3/59-7/27/59
VAH-6/CVG-9	A3D-2	USS *Ranger* (CVA-61)	NG x	2/6/60-8/30/60
VAH-6/CVG-9	A3D-2	USS *Ranger* (CVA-61)	NG x	8/11/61-3/8/62
VAH-6/CVG-9	A3D-2	USS *Ranger* (CVA-61)	NG x	11/9/62-6/14/63
VAH-8/CVG-2	A3D-2	USS *Midway* (CVA-41)	ZD 4xx	8/16/58-3/12/59
VAH-8/CVG-2	A3D-2	USS *Midway* (CVA-41)	ZD 4xx	8/15/59-3/25/60
VAH-8/CVG-2	A3D-2	USS *Midway* (CVA-41)	NE 6xx	2/15/61-9/28/61
VAH-8/CVG-2	A3D-2	USS *Midway* (CVA-41)	NE 6xx	4/6/62-10/20/62
VAH-8/CVW-2	A-3B	USS *Midway* (CVA-41)	NE 6xx	11/8/63-5/26/64
VAH-10/CVG-14	A-3B	USS *Constellation* (CVA-64)	NK 2xx	2/63-9/63
VAH-13/CVG-11	A-3B	USS *Kitty Hawk* (CVA-63)	GP/NH 6xx	9/13/62-4/2/63
VAH-13/CVW-11	A-3B	USS *Kitty Hawk* (CVA-63)	NH 6xx	10/13/63-6/16/64
VAQ-130 Det 3/CVW-19	KA/EKA-3B	USS *Oriskany* (CVA-34)	NM 61x	10/18/73-6/5/74
VAQ-130 Det 4/CVW-2	KA/EKA-3B	USS *Ranger* (CVA-61)	NE 61x	5/7/74-10/18/74

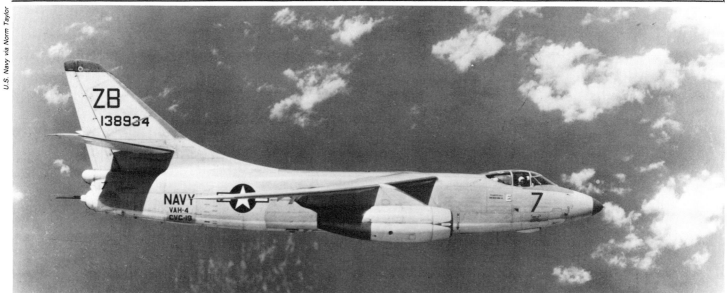

Detachments from Heavy Attack Squadron Four made three deployments aboard USS "Bon Homme Richard" (CVA-31) while assigned to CVG-19. This photograph of an early A3D-2 was taken at the time of the first of these deployments, when Det Echo went on a cruise in the WestPac from November 21, 1959, until May 14, 1960. BuNo 138934 was lost in March, 1963, while operated by VAH-13 from NAS Whidbey Island.

U.S. Navy via Norm Taylor

SKYWARRIOR HATRON & TACELRON DEPLOYMENTS
SOUTHEAST ASIA WAR
AUGUST 2, 1964 – AUGUST 15, 1973

Squadron/Air Wing	Aircraft	Carrier	Tail Code & Modex	Deployment Dates
VAH-2 Det M/CVW-9	A-3B	USS *Ranger* (CVA-61)	NL 81x	8/5/64–5/6/65
VAH-2/CVW-15	A-3B	USS *Coral Sea* (CVA-43)	NL 6xx	12/7/64–11/1/65
VAH-2 Det F/CVW-14	A-3B	USS *Ranger* (CVA-61)	ZA 81x	12/10/65–8/25/66
VAH-2 Det A/CVW-2	A-3B	USS *Coral Sea* (CVA-43)	ZA 6xx	7/29/66–2/23/67
VAH-2 Det M/CVW-9	A-3B	USS *Enterprise* (CVAN-65)	ZA 11x	11/19/66–7/6/67
VAH-2 Det 43/CVW-15	KA-3B	USS *Coral Sea* (CVA-43)	NL 6xx	7/26/67–4/6/68
VAH-2 Det 61/CVW-2	KA-3B	USS *Ranger* (CVA-61)	ZA 6xx	11/4/67–5/25/68
VAH-2 Det 65/CVW-9	KA-3B	USS *Enterprise* (CVAN-65)	ZA x	1/3/68–7/18/68
VAH-2 Det 64/CVW-14	KA-3B	USS *Constellation* (CVA-64)	NK 11x	5/29/68–10/31/68
VAH-4 Det E/CVW-19	A-3B	USS *Bon Homme Richard* (CVA-31)	ZB xx	1/28/64–11/21/64
VAH-4 Det B/CVW-5	A-3B	USS *Ticonderoga* (CVA-14)	ZB x	4/14/64–12/15/64
VAH-4 Det L/CVW-21	A-3B	USS *Hancock* (CVA-19)	ZB x	10/21/64–5/29/65
VAH-4 Det G/CVW-16	A-3B	USS *Oriskany* (CVA-34)	ZB xx	4/5/65–12/16/65
VAH-4 Det 62/CVW-7	A-3B	USS *Independence* (CVA-62)	ZB x/AG 8xx	5/10/65–12/13/65
VAH-4 Det B/CVW-5	A-3B	USS *Ticonderoga* (CVA-14)	ZB 8xx	9/28/65–5/13/66
VAH-4 Det C/CVW-11	A-3B	USS *Kitty Hawk* (CVA-63)	ZB x	10/19/65–6/13/66
VAH-4 Det M/CVW-9	A-3B	USS *Enterprise* (CVAN-65)	ZB 61x	10/26/65–6/21/66
VAH-4 Det G/CVW-16	A-3B	USS *Oriskany* (CVA-34)	ZB 61x	5/26/66–11/16/66
VAH-4 Det E/CVW-19	A-3B	USS *Ticonderoga* (CVA-14)	ZB 1x	10/15/66–5/29/67
VAH-4 Det C/CVW-11	A-3B	USS *Kitty Hawk* (CVA-63)	ZB x	11/5/66–6/20/67
VAH-4 Det B/CVW-5	A-3B	USS *Hancock* (CVA-19)	ZB x	1/5/67–7/22/67
VAH-4 Det L/CVW-21	A-3B	USS *Bon Homme Richard* (CVA-31)	ZB 8xx	1/26/67–8/25/67
VAH-4 Det G/CVW-16	KA-3B	USS *Oriskany* (CVA-34)	AH 61x	6/26/67–1/31/68
VAH-4 Det 63/CVW-11	KA-3B	USS *Kitty Hawk* (CVA-63)	ZB x	11/18/67–6/28/68
VAH-4 Det 14/CVW-19	KA-3B	USS *Ticonderoga* (CVA-14)	ZB 1x	12/28/67–8/17/68
VAH-8/CVW-2	A-3B	USS *Midway* (CVA-41)	NE 6xx	3/6/65–11/23/65
VAH-8/CVW-15	A-3B	USS *Constellation* (CVA-64)	NL 6xx	5/12/66–12/3/66
VAH-8/CVW-14	A-3B/KA-3B	USS *Constellation* (CVA-64)	NK 1xx	4/29/67–12/4/67
VAH-10/CVW-14	A-3B	USS *Constellation* (CVA-64)	NK 1xx	5/5/64–2/1/65
VAH-10 Det 42/CVW-1	A-3B	USS *Franklin D. Roosevelt* (CVA-42)	AB 6xx	6/21/66–2/21/67
VAH-10 Det 59/CVW-17	KA-3B	USS *Forrestal* (CVA-59)	AA 61x	6/6/67–9/14/67
VAH-10 Det 66/CVW-6	KA-3B	USS *America* (CVA-66)	AE 01x	4/10/68–12/16/68
VAH-10 Det 43/CVW-15	KA-3B	USS *Coral Sea* (CVA-43)	NL 60x	9/7/68–4/18/69
VAH-10 Det 61/CVW-2	KA-3B	USS *Ranger* (CVA-61)	NE 61x	10/26/68–5/17/69
VAH-10 Det 19/CVW-21	KA-3B	USS *Hancock* (CVA-19)	NP 61x	8/2/69–4/15/70
VAH-10 Det 38/CVW-8	KA-3B	USS *Shangri-La* (CVS-38)	AJ 61x	3/5/70–8/31/70

A3D-2 of VAH-2, seen during an aborted landing on January 2, 1961, during its tour with the USS "Coral Sea" (CVA-43). Airbrake, tail wheel, tail hook, flap, and leading edge slat positions are noteworthy.

A-3B, BuNo 147655, of VAH-2 and stationed aboard the USS "Coral Sea" (CVA-43), during 1965, while operating over the Gulf of Tonkin. Completed mission marks are readily discernible on the fuselage side, just under the unit patch.

A KA-3B (BuNo 147657) comes in for recovery aboard USS "Constellation" (CVA-64). This photograph was taken in the Gulf of Tonkin on August 14, 1967, when the "Fireballers" of Heavy Eight were making their last war cruise.

A-3B, BuNo 147655, of VAH-2, onboard the USS "Coral Sea" (CVA-43) while in the Gulf of Tonkin during the spring of 1965. Visible on the fuselage side is the VAH-2 unit patch and five mission marks (which appear to be small bombs painted in black).

Deployments Cont.

Squadron/Wing	Aircraft	Ship	Code	Dates
VAQ-129 Det 38/CVW-8	KA-3B	USS *Shangri-La* (CVS-38)	AJ 61x	9/1/70–12/17/70
VAQ-129 Det 19/CVW-21	EKA-3B	USS *Hancock* (CVA-19)	NP 61x	10/22/70–6/2/71
VAQ-130 Det 43/CVW-15	EKA-3B	USS *Coral Sea* (CVA-43)	VR 01x	9/7/68–4/18/69
VAQ-130 Det 61/CVW-2	EKA-3B	USS *Ranger* (CVA-61)	NE 61x	10/26/68–5/17/69
VAQ-130 Det 14/CVW-16	EKA-3B	USS *Ticonderoga* (CVA-14)	AH 61x	2/1/69–9/18/69
VAQ-130 Det 31/CVW-5	KA/EKA-3B	USS *Bon Homme Richard* (CVA-31)	NF 61x	3/18/69–10/29/69
VAQ-130 Det 34/CVW-19	EKA-3B	USS *Oriskany* (CVA-34)	NM 61x	4/16/69–11/17/69
VAQ-130 Det 31/CVW-5	EKA-3B	USS *Bon Homme Richard* (CVA-31)	NF 61x	4/2/70–11/12/70
VAQ-130 Det 1/CVW-19	EKA-3B	USS *Oriskany* (CVA-34)	NM 61x	5/16/70–12/10/70
VAQ-130 Det 2/CVW-5	EKA-3B	USS *Midway* (CVA-41)	NF 61x	4/16/71–11/6/71
VAQ-130 Det 3/CVW-19	EKA-3B	USS *Oriskany* (CVA-34)	NM 61x	5/14/71–12/18/71
VAQ-130 Det 4/CVW-14	EKA-3B	USS *Enterprise* (CVAN-65)	NK 61x	6/11/71–2/12/72
VAQ-130 Det 1/CVW-9	EKA-3B	USS *Constellation* (CVA-64)	NG 61x	10/1/71–7/1/72
VAQ-130 Det 2/CVW-5	EKA-3B	USS *Midway* (CVA-41)	NF 61x	4/10/72–3/3/73
VAQ-130 Det 3/CVW-19	EKA-3B	USS *Oriskany* (CVA-34)	NM 61x	6/5/72–3/30/73
VAQ-130 Det 4/CVW-2	EKA-3B	USS *Ranger* (CVA-61)	NE 61x	11/16/72–6/22/73
VAQ-130 Det 5/CVW-21	KA/EKA-3B	USS *Hancock* (CVA-19)	NP 61x	5/8/73–1/8/74

(This detachment had started the cruise as VAQ-135 Det 5 but was redesignated VAQ-130 Det 5 on 8/25/73.)

Squadron/Wing	Aircraft	Ship	Code	Dates
VAQ-131/CVW-11	KA/EKA-3B	USS *Kitty Hawk* (CVA-63)	NH 61x	12/30/68–9/4/69
VAQ-132 Det 64/CVW-14	KA-3B	USS *Constellation* (CVA-64)	NK 11x	11/1/68–1/31/69
VAQ-132/CVW-9	KA/EKA-3B	USS *Enterprise* (CVAN-65)	NG 61x	1/6/69–7/2/69
VAQ-132/CVW-9	KA/EKA-3B	USS *America* (CVA-66)	NG 61x	4/10/70–12/21/70
VAQ-133/CVW-14	KA/EKA-3B	USS *Constellation* (CVA-64)	NK 61x	8/11/69–5/8/70
VAQ-133/CVW-11	KA/EKA-3B	USS *Kitty Hawk* (CVA-63)	NH 61x	11/6/70–7/17/71
VAQ-134/CVW-2	KA/EKA-3B	USS *Ranger* (CVA-61)	NE 61x	10/14/69–6/1/70
VAQ-134/CVW-2	KA/EKA-3B	USS *Ranger* (CVA-61)	NE 61x	10/27/70–6/17/71
VAQ-135/CVW-15	KA/EKA-3B	USS *Coral Sea* (CVA-43)	NL 61x	9/23/69–7/1/70
VAQ-135 Det 3/CVW-15	EKA-3B	USS *Coral Sea* (CVA-43)	NL 61x	11/12/71–7/17/72
VAQ-135 Det 5/CVW-21	EKA-3B	USS *Hancock* (CVA-19)	NP 61x	1/7/72–10/3/72
VAQ-135/CVW-11	EKA-3B	USS *Kitty Hawk* (CVA-63)	NH 61x	2/17/72–11/28/72
VAQ-135 Det 3/CVW-15	EKA-3B	USS *Coral Sea* (CVA-43)	NL 61x	3/9/73–11/8/73
VAQ-135 Det 5/CVW-21	KA/EKA-3B	USS *Hancock* (CVA-19)	NP 61x	5/8/73–1/8/74

(During this cruise, the detachment was redesignated VAQ-130 Det 5 on 8/25/73.)

Squadron/Wing	Aircraft	Ship	Code	Dates
VAW-13 Det 61/CVW-2	EKA-3B	USS *Ranger* (CVA-61)	VR 72x	11/4/67–5/25/68
VAW-13 Det 65/CVW-9	EKA-3B	USS *Enterprise* (CVAN-65)	ZA x	1/3/68–7/18/68
VAW-13 Det 31/CVW-5	EKA-3B	USS *Bon Homme Richard* (CVA-31)	NF 03x	1/27/68–10/10/68
VAW-13 Det 66/CVW-6	EKA-3B	USS *America* (CVA-66)	AE 71x	4/10/68–12/16/68
VAW-13 Det 64/CVW-14	EKA-3B	USS *Constellation* (CVA-64)	NK 11x	5/29/68–10/31/68
VAW-13 Det 19/CVW-21	EKA-3B	USS *Hancock* (CVA-19)	NP 84x	7/18/68–3/3/69

Taken in June, 1971, this photo shows an EKA-3B (BuNo 142654) of VAQ-134 shortly after this TACELRON returned to NAS Alameda at the end of its last war cruise aboard USS "Coral Sea" (CVA-42).

On November 6, 1970, two weeks after this photograph was taken at NAS Alameda, BuNo 147665, an EKA-3B of VAQ-133, departed for Yankee Station aboard USS "Kitty Hawk" (CVA-63) for a last cruise before converting to EA-6Bs.

Assigned to CVW-21, VAQ-129 departed for Southeast Asia aboard USS "Hancock" (CVA-19) on October 22, 1970. One of its EKA-3Bs is seen here at NAS Alameda prior to this war cruise.

An EKA-3B (BuNo 142660) of VAQ-130 Det 3 operating over the Gulf of Tonkin in June, 1970, while assigned to CVW-5 aboard the USS "Bon Homme Richard" (CVA-31).

SKYWARRIOR COMBAT LOSSES DURING THE SOUTHEAST ASIA WAR

DATE	MODEL	BuNo	SQUADRON/ AIR WING	CARRIER OR BASE	CAUSE/LOCATION	FATE OF CREW
04/12/66	A-3B	142653	VAH-4/CVW-11	USS *Kitty Hawk* (CVA-63)	Probably MiG China Sea during ferry flight from NAS Cubi Point	KIA
06/13/66	RA-3B	144842	VAP-61	NAS Cubi Point	AAA North Vietnam during night reconnaissance in RP-2	KIA
03/08/67	A-3B	144627	VAH-4/CVW-11	USS *Kitty Hawk* (CVA-63)	Unknown North Vietnam during night mining of Kien Giang River	KIA
08/25/67	RA-3B	144835	VAP-61	NAS Cubi Point	Unknown North Vietnam during night reconnaissance	KIA
10/14/67	RA-3B	144844	VAP-61	NAS Cubi Point	AAA	2 KIA, 1 recovered
01/01/68	RA-3B	144847	VAP-61	NAS Cubi Point	Unknown North Vietnam during night reconnaissance in RP-2	KIA

SKYWARRIOR OPERATIONAL LOSSES DURING THE SOUTHEAST ASIA WAR

DATE	MODEL	BuNo	SQUADRON/ AIR WING	CARRIER OR BASE	CAUSE/LOCATION	FATE OF CREW
12/27/64	A-3B	142250	VAH-4/CVW-21	USS *Hancock* (CVA-19)	Control system failure & fire	3 recovered, 1 killed
02/24/65	A-3B	147664	VAH-2/CVW-15	USS *Coral Sea* (CVA-43)	Fuel transfer malfunction	3 recovered, 1 killed
05/25/65	A-3B	138947	VAH-4/CVW-16	USS *Oriskany* (CVA-34)	Structural failure on launch	4 recovered
04/01/66	A-3B	142665	VAH-4/CVW-9	USS *Enterprise* (CVAN-65)	Nose wheel collapse on launch	3 killed
10/02/66	A-3B	142633	VAH-2/CVW-2	USS *Coral Sea* (CVA-43)	Shed bridle on launch	4 recovered
06/16/67	RA-3B	144828	VAP-61	NAS Cubi Point	Fire & loss of control	3 recovered
07/28/67	KA-3B	142658	VAH-4/CVW-16	USS *Oriskany* (CVA-34)	Double engine failure	1 recovered, 2 killed
10/21/67	KA-3B	142655	VAH-4/CVW-16	USS *Oriskany* (CVA-34)	JATO bottles ignited on launch	4 recovered
11/03/67	KA-3B	147653	VAH-8/CVW-14	USS *Constellation* (CVA-64)	Premature bridle separation	3 killed
02/17/69	KA-3B	138943	VAH-10/CVW-15	USS *Coral Sea* (CVA-43)	Probable pilot fatigue	3 killed
08/08/69	RA-3B	144826	VAP-61	DaNang AB	Erratic fuel flow & flame-out	4 recovered
05/16/70	EKA-3B	142657	VAQ-135/CVW-15	NAS Cubi Point	Contact lost on ferry to CVA-43	1 killed, 2 missing
07/04/70	EKA-3B	142400	VAQ-132/CVW-9	USS *America* (CVA-66)	Drag chute opened after bolter	3 recovered
06/18/71	EKA-3B	147649	VAQ-130	DaNang AB	Control failure	3 killed
06/21/73	EKA-3B	142634	VAQ-130/CVW-2	USS *Ranger* (CVA-61)	Launch accident	3 killed

SKYWARRIOR ASSIGNMENT DECEMBER 1985

KA-3B	138925	VAQ-33	**EA-3B** (cont.)	144852	VQ-2	**ERA-3B** (cont.)	144846	VAQ-34
	138944	VAQ-34		144854	VQ-1		146446	VAQ-33
	142650	VAQ-33		146448	VQ-2		146447	VAQ-33
	142662	VAK-308		146451	VQ-1	**NRA-3B**	1442256	Bailed to Westinghouse
	142664	VAK-208		146452	VQ-1		142667	PMTC
	147648	VAK-308		146453	VQ-2		144825	PMTC
	147655	VAK-208		146454	VQ-2		144840	PMTC
	147656	VAK-308		146455	VQ-2	**TNRA-3B**	144834	VQ-2
	147657	VAK-208		146457	VQ-1	**TA-3B**	144856	VAQ-33
	147663	VAK-208		146459	VQ-1		144858	VAQ-33
	147665	VAK-208	**RA-3B**	144833	PMTC		144859	VAQ-33
	147666	VAK-308		144843	Bailed to Raytheon		144862	VAQ-33
	147667	VAK-308					144866	VAQ-33
NA-3B	138938	PMTC	**ERA-3B**	142668	VAQ-34		144867	Bailed to Grumman
	142630	NWC		144827	VAQ-33			
EA-3B	142671	VQ-1		144832	VAQ-33	**TA-3B** (VIP-conf.)	144857	CFLSW Det
	144849	VQ-1		144838	VAQ-34		144865	CFLSW Det
	144850	VQ-2		144841	VAQ-34			

Remark: On February 6, 1986, BuNo 144865 was brought to NARF Alameda for SDLM (Standard Depot Level Maintenance). It is anticipated that this TA-3B will not be returned to CFLSW but will be assigned to VQ-1 as a trainer.

JQ 17, an EA-3B of VQ-2 in flight off the coast of Spain in July, 1980. EA-3Bs from this unit have seen considerable use in the Mediterranean and, notably, have kept their electronic ears attuned to Libyan communications.

BuNo 144857, one of the last two VIP-configured TA-3Bs assigned to CFLSW Det at Andrews AFB. At the end of 1985, eight of the 12 A3D-2Ts which had been accepted between April, 1959, and July, 1960, were still operational.

Chapter 7
Unit Histories

Photographed on August 5, 1960, one day after CVG-7 left Florida for a Med cruise, this A3D-1 was being preflighted prior to launch as its wings were hydraulically unfolded. This was the second of four operational cruises which VAH-1 made before transitioning to North American A3J-1s during the fall of 1962. This A3D-1 (BuNo 135436) was salvaged at NARF Alameda in January, 1968.

Since 1956, thirteen Heavy Attack Squadrons (VAHs or HATRONs), two Aerial Refueling Squadrons (VAKs or AERREFRONs), two Heavy Photographic Squadrons (VAPs or HATPHOTORONs), nine Tactical Electronic Warfare Squadrons (VAQs or TACELRONs), one Carrier Airborne Early Warning Squadron (VAW or CARAEWRON), one Composite Photographic Squadron (VCP or PHOTOCOMPRON), two Fleet Air Reconnaissance Squadrons (VQs or FAIRECONs), and one Fleet Transport Squadron (VR or TRANSRON) have flown *Skywarriors*. In addition, A-3s have been operated by a Naval Air Reserve Unit (NARU) as well as by and at several specialized naval units and facilities (ARF, COMFLELOGSUPPWING, FEWSG, NADC, NAPOG, NASWF/NWEF, NATC, NATF, NMC/PMTC, and NOTS/NWC). In this chapter, *Skywarrior* squadron histories are followed by synopses of *Skywarrior* operations by other naval units.

HEAVY ATTACK SQUADRON ONE
VAH-1 "TIGERS"

Commissioned at NAS Jacksonville, Florida, on November 1, 1955, as the heir of Patrol Squadron Sixteen (VP-16), VAH-1 gained the distinction of being the first operational unit to receive *Skywarriors* when five A3D-1s were ferried from NAS Patuxent River, Maryland, five months later. After initial training at its home base, VAH-1 completed carrier qualifications aboard the USS *Forrestal* (CVA-59) in October, 1956, and was thus ready to bolster the Sixth Fleet during the Suez Crisis in November. This first cruise, which lasted only one month, was followed by two full-length cruises aboard CVA-59 in 1957-58, one in the Mediterranean and one in the North Atlantic.

Moving to NAS Sanford, Florida, in January, 1959, HATRON One recorded the first operational launch and arrested landing aboard the USS *Independence* (CVA-62) in May of that year. Two Mediterranean cruises were made aboard this carrier in 1959-60 and in 1961. Between these two cruises, VAH-1 also provided inflight refueling for Project LANA, the Bendix Trophy transcontinental speed record set by McDonnell F4H-1s to commemorate the 50th Anniversary of U.S. Naval Aviation.

Returning from its fourth Mediterranean cruise in August, 1962, after setting a utilization record of 885 flights hours in 17 days aboard *Independence*, VAH-1 began its transition to North American A3J-1s. VAH-1 disposed of its last A-3Bs prior to the end of 1962, and the unit became Reconnaissance Heavy Attack Squadron One on September 1, 1964 after transitioning in the RA-5C.

This pair of A3D-2s (BuNos 138916, AG 600, and 138918, AG 608) was photographed on May 11, 1962, during VAH-1's third cruise aboard the USS "Independence" (CVA-62). The open bomb bays with their deployed spoiler, dovetail fairing and flat radome are well in evidence in this view.

BuNo 138917, an A3D-2 assigned to VAH-2 Det Bravo, was photographed in May 1957, during carquals aboard USS "Bon Homme Richard" (CVA-31). By deploying aboard a CVA-19 class angled-deck carrier, this Det vindicated Ed Heinemann's claims that a heavy attack jet could be designed to operate from relatively small carriers.

BuNos 147650 and 147666, two war veterans from VAH-2, photographed over the Mekong Delta in 1965. At that time, NL 606 had flown 11 bombing sorties and NL 601 had flown 22, as evidenced by the black bombs painted on the forward fuselage beneath the refueling probe.

HEAVY ATTACK SQUADRON TWO
VAH-2 "ROYAL RAMPANTS"

This HATRON was formed from VP-29 to become PacFlt's first heavy jet attack squadron. Commissioned at NAS Whidbey Island, Washington, on November 1, 1955, VAH-2 was inititally equipped with Lockheed TV-2 jet trainers and with Lockheed P2V-5Fs and P2V-3Bs fitted with radar and navigation/bombing systems similar to those of their forthcoming *Skywarriors*. During 1956, a few Douglas F3D-2Ts were added. As pilot and crew training progressed satisfactorily, personnel from VAH-2 joined VAH-1 crews to participate in the A3D Fleet Indoctrination Program at NAS Patuxent River, Maryland. Temporarily homeported at NAS North Island, California, from January, 1956, until December, 1957, the unit received its first A3D-1 in May, 1956, and its first A3D-2 two months later. Initial A3D-2 carquals began in May, 1957, aboard the USS *Bon Homme Richard* (CVA-31), with Lt. Ed Mitchell—a future astronaut—becoming the first VAH-2 pilot to complete day carquals; the first night qualification was made by Cdr. H. L. Salyer.

While aboard the USS *Bon Homme Richard*, between July and December, 1957, VAH-2 Det. Bravo became the first A3D unit to deploy to WestPac; it was followed by Det. Mike aboard the USS *Ticonderoga* (CVA-14) (Sep '57-Apr '58). Homeported at NAS Whidbey Island, Washington, from December, 1957, VAH-2 crews and support personnel then deployed several times aboard these two carriers; later, they also embarked aboard the USS *Coral Sea* (CVA-43) and *Ranger* (CVA-61) (initially in Sept. '60 and Oct. '64, respectively).

VAH-2's first war cruise was made aboard the USS *Coral Sea* (December 7, 1964-November 1965), with the unit being credited with the first A-3 bombing sorties (March 29, 1965). During the Southeast Asia War, VAH-2 detachments made eight more deployments to the Gulf of Tonkin. In the course of these operations, VAH-2 primarily provided air refueling for the fleet with A-3Bs configured as tankers and later with KA-3Bs. Finally, as combined tanking/electronic warfare came into being, VAH-2 was redesignated VAQ-132 on November 1, 1968.

This TA-3B (BuNo 144859) of RVAH-3, which was photographed at NAS Key West on August 11, 1975, was one of the few TA-3Bs fitted with the blunt nose during overhaul at NARF Alameda. Note the inboard slat section which was characteristic of aircraft with the CLE wing.

HEAVY ATTACK SQUADRON THREE
VAH-3 "SEA DRAGONS"

On June 15, 1956, VP-34 was disestablished and VAH-3 was commissioned at NAS Jacksonville, Florida, as an operational A3D-1 squadron. After completing its carquals aboard the USS *Franklin D. Roosevelt* (CVA-42), VAH-3 deployed in July, 1957, for its only cruise with *Skywarriors*. In June, 1958, three months after returning from this Mediterranean cruise, the squadron moved to NAS Sanford, Florida, to be merged with the Atlantic Fleet's Heavy Attack Training Unit. In its new role as a RAG Squadron, VAH-3 took over HATULANT's Douglas R4D-7s and Grumman F9F-8Ts, and in late 1959 added A3D-2Ts to its mixed complement of A3D-1s and A3D-2s.

During 1960, VAH-3 was given the additional duty of training crews for the North American *Vigilante*, and the first A3J-1s were received in June, 1961. Training of A-3 crews then became progressively less important, with the last class of *Skywarrior* pilots and bombardiers graduating in January, 1964. As the RA-5C RAG squadron, the unit was redesignated RVAH-3 six months later, moved to NAS Albany, Georgia, in May, 1968, and to NAS Key West, Florida, in January, 1974. RVAH-3 continued operating TA-3Bs until disestablished in August, 1979.

After being replaced in service with deployable squadrons of more advanced and better equipped A3D-2s, most A3D-1s ended their lives with various Navy training squadrons and there served as instructional tools for fledgling A-3 crews. Typical of "Skywarriors" relegated to this role is A3D-1, BuNo 135419, of VAH-3.

HEAVY ATTACK SQUADRON FOUR
VAH-4 "FOURRUNNERS"

HATRON Four, formerly VP-57, came into being on July 1, 1956 as the second *Skywarrior* squadron in PacFlt. Following a 14-month training program (during which one of its aircraft set a record of 4 hr. 29 min. 50 sec. between the West Coast and Hawaii), VAH-4 first deployed to WestPac in February, 1958, when Detachment Delta went aboard the USS *Hancock* (CVA-19). VAH-4's peacetime activities were also marked by its contribution of A3D-2 tankers to Project LANA in May, 1961.

Remaining homeported at NAS Whidbey Island, Washington, throughout its 12-year existence, HATRON Four never deployed as a full squadron aboard a single carrier. During the early years, the squadron's deployment history was unique as its A-3B detachments only went aboard the smaller attack carriers. During the Southeast Asia War, however, VAH-4 detachments also embarked aboard larger carriers (initially aboard the USS *Independence* (CVA-62) in 1965) and, as there was a shortage of tanker-configured A-3Bs to support Task Force 77 operations, in 1966 the squadron also sent a detachment to operate from NAS Cubi Point in the Philippines and Da Nang AB in South Vietnam. Altogether, VAH-4 made 16 war deployments aboard carriers, more than any other *Skywarrior* squadron.

VAH-4 last deployed KA-3Bs aboard the USS *Ticonderoga* (CVA-14) in December, 1967. It returned to Whidbey Island eight months later. By then it had been decided that on November 1, 1968, VAH-4 would be reorganized as VAQ-131 to provide tanking and electronic support.

With the boost of twelve 4,500 lb. thrust Aerojet JATO bottles, "Skywarriors" could lift off in the length of a football field as demonstrated on October 16, 1958, by this A3D-2 (BuNo 138912) of VAH-4. Due to the angle at which the bottles were attached, the maximum vectored thrust was 45,000 lbs.

A-3B (BuNo 142239) of VAH-4 Det Echo was photographed at Elmendorf AFB, Alaska, on June 27, 1967, when it was ferried back to CONUS after making a war cruise aboard USS "Ticonderoga" (CVA-14) from October 15, 1966, to May 29, 1967 as part of CVW-19.

Photographed over North Vietnam, ZB 896 was one of the A-3Bs of VAH-4 Det Bravo which were aboard USS "Ticonderoga" (CVA-14) from September 28, 1965 until May 13, 1966. VAH-4 detachments made 16 war deployments aboard carriers, more than any other "Skywarrior" squadron.

A quartet of A3D-2s from VAH-5 over Florida on November 20, 1957, two months after VAH-5 received its first "Skywarriors" to become the fourth squadron in Heavy Attack Wing One to be so equipped. Formation flying in the A-3 was not difficult, but because of the size of the aircraft slow reaction times had to be taken into consideration.

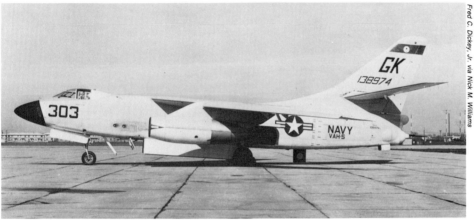

A3D-2, BuNo 138974, from VAH-5 at NAS Glenview during a temporary stopover in Illinois. After making five Mediterranean deployments, VAH-5 transitioned into the North American RA-5C and was redesignated RVAH-5 during May, 1964.

HEAVY ATTACK SQUADRON FIVE
VAH-5 "SAVAGE SONS"

Originating at NAS Moffett Field, California, in September, 1948, as Composite Squadron Five (VC-5), this squadron was a pioneer of heavy carrier aircraft operations. Initially equipped with Lockheed P2V-3Cs, VC-5 converted to North American AJ-1s during the spring of 1950 and flew *Savages* for seven years. During this period, it moved to NAS Norfolk, Virginia, in late 1950, to NAS Jacksonville, Florida, in 1952, and to NAS Sanford, Florida, in 1955.

Redesignated Heavy Attack Squadron Five on February 3, 1956, the Savage Sons had to wait 18 months to receive their first *Skywarriors*. With A3Ds, VAH-5 won several awards while compiling a brilliant record and made five Med deployments. Returning to NAS Sanford in March, 1963, at the end of a Mediterranean cruise, the squadron began transitioning to North American RA-5Cs and was redesignated RVAH-5 in May, 1964, thus ending its five-year association with the *Skywarrior*.

A VAH-5 A-3B being prepared for launch from the forward port catapult of the USS "Forrestal" (CVA-59) while operating with the Sixth Fleet in the Mediterranean on October 19, 1962. Catapult launches of the heavy A-3 required special considerations on behalf of carrier deck crews.

BuNo 146653, an A-3B of VAH-6, was photographed at NAS Alameda on July 20, 1963, five weeks after this squadron completed its fourth and last WestPac deployment aboard USS "Ranger" (CVA-61). VAH-6 then made a Med cruise aboard USS "Forrestal" (CVA-59) prior to converting to RA-5Cs.

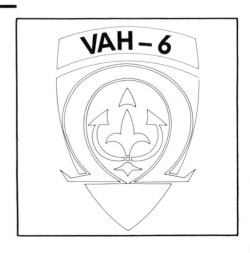

HEAVY ATTACK SQUADRON SIX
VAH-6 "FLEURS"

In January, 1950, Composite Squadron Six was formed at NAS Moffett Field, California, as the Navy's second nuclear attack squadron. Like its sister unit, VC-5, the squadron flew *Neptunes* and *Savages* for many years. It moved to NAS North Island, California, in June, 1952, and was redesignated VAH-6 on July 1, 1956. In 1958, after moving to NAS Whidbey Island, Washington, VAH-6 converted to *Skywarriors*. As part of CVG-6/CVW-6 the squadron then made several WestPac deployments aboard the USS *Ranger* (CVA-61) prior to transferring to CVW-8 for operations in the Mediterranean aboard the USS *Forrestal* (CVA-59). Transferred to NAS Sanford, Florida, the squadron was redesignated RVAH-6 in September, 1965, and commenced its transition to the RA-5C *Vigilante*.

William T. Larkins

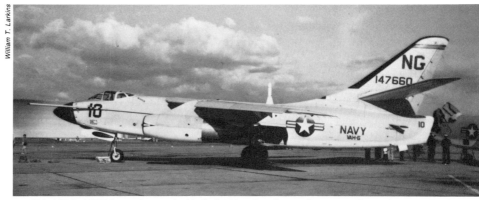

This A3D-2 of VAH-6 was photographed in April, 1961, when the squadron was working up in preparation for its third deployment aboard USS "Ranger" (CVA-61). BuNo 147660 was last operated by VAK-308 prior to being struck off at NARF Alameda in August, 1963, due to excessive corrosion.

R. C. Seely via David W. Menard

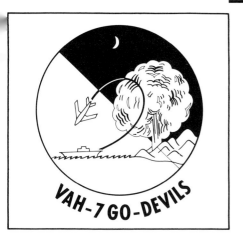

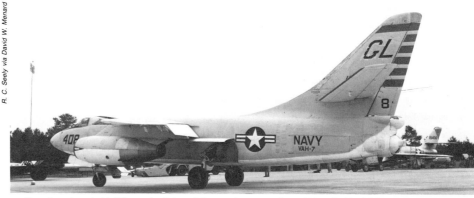

Though equipped with "Skywarriors" for a little over three years from 1958 until 1961, VAH-7 made only a partial deployment with the type when some of its A3D-2s and their crews supplemented VAH-1 aboard the USS "Independence" (CVA-62).

HEAVY ATTACK SQUADRON SEVEN
VAH-7 "PEACEMAKERS OF THE FLEET"

Commissioned at NAS Moffett Field, California, in October, 1950, Composite Squadron Seven moved to NAS Patuxent River, Maryland, in 1951, and to NAS Sanford, Florida, in 1955. Redesignated VAH-7 on July 1, 1955, the squadron exchanged its *Savages* for *Skywarriors* in early 1958. However, VAH-7's association with the Douglas heavy attack aircraft lasted only a little over three years (during which only a partial deployment was made when some VAH-7 aircraft and crews supplemented VAH-1 aboard the USS *Independence*, CVA-62) as it began its transition to North American A3J-1s in August, 1961. After transitioning to the RA-5C, the Peacemakers became RVAH-7 on December 1, 1964.

C. Seeley via William T. Larkins

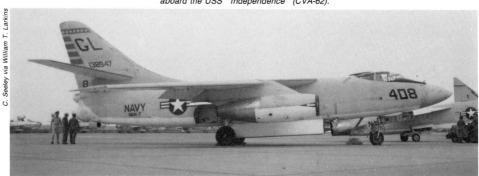

Seen here in August, 1960, while it served with VAH-7, BuNo 138947 later became the third "Skywarrior" to be lost in an operational accident during the Southeast Asia War (a structural failure while being launched from the USS "Oriskany" [CVA-34] on May 25, 1965).

HEAVY ATTACK SQUADRON EIGHT
VAH-8 "FIREBALLERS"

In existence for less than 11 years, VAH-8

U.S. Navy via Harvey Herzog

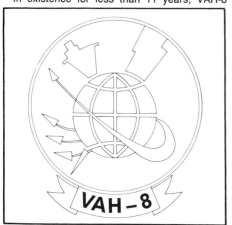

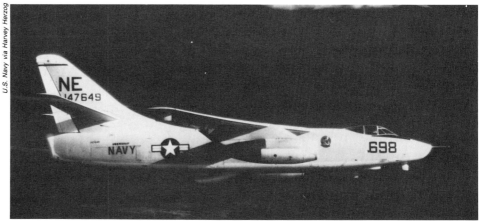

This tanker-configured A-3B of VAH-8 was photographed over North Vietnam on August 19, 1965. At that time VAH-5 was assigned to CVW-2 for operations aboard the USS "Midway" (CVA-41). BuNo 147649 was lost off Da Nang on June 18, 1971, while flown by VAQ-130.

Bearing the tail code ZD, indicating that it belonged to the fourth squadron (VAH-8) in Heavy Attack Wing Two, this A3D-2 (BuNo 142240) was photographed in March, 1958, after trapping aboard the USS "Midway" (CVA-41). The aircraft has rolled aft and the cross deck pendant has fallen off the hook, freeing the aircraft to taxi forward.

nevertheless became one of the foremost proponents of the *Skywarrior*. Commissioned as VAH-8 at NAS Whidbey Island, Washington, on May 1, 1957, the squadron received its first A3D-2 three months later and departed for its first WestPac deployment aboard the USS *Midway* (CVA-41) in August, 1958. At the end of its next WestPac cruise in March, 1960, nine of its A3D-2s accomplished a "first" when they were launched from the *Midway* some 2100 miles west of Hawaii; refueling at NAS Barbers Point, Hawaii, and NAS Alameda, California, these aircraft flew over 5500 miles in 10.9 hours to reach their homeport at NAS Whidbey Island.

During the Southeast Asia War, VAH-8 made three combat cruises (one aboard the *Midway* and two aboard the *Constellation*, CVA-64) prior to being disestablished on January 17, 1968.

This A3D-2 (BuNo 138941) was photographed during VAH-9's first cruise aboard the USS "Saratoga" (CVA-60) which lasted from February 1 until October 1, 1958. It is seen being towed to a new spot prior to the start of an evolution. Note the bracing strut holding the folded wing panels.

VAH-9 was one of three "Skywarrior" squadrons which refueled F4H-1F "Phantom IIs" during their record-setting transcontinental flight on May 24, 1961, as part of Operation LANA. The tanker-configured A3D-2s of VAH-9 transferred fuel to the F4H-1Fs of VF-101 over Pennsylvania.

HEAVY ATTACK SQUADRON NINE
VAH-9 "HOOT OWLS"

The squadron came into being in January, 1953, when it was commissioned as Composite Squadron Nine at NAS Sanford, Florida. Flying *Savages*, VC-9 conducted the Navy's first inflight refueling operations while deployed aboard the USS *Midway* (CVA-41) in 1953. Redesignated VAH-9 on November 1, 1955, the unit received its first *Skywarriors* 14 months later. Thereafter VAH-9 flew A3Ds for eight years, making six Med deployments aboard the USS *Saratoga* (CVA-60). Transition to the RA-5C began in April, 1964, the unit's designation was changed to RVAH-9 in June, and the last A-3B was transferred out on August 4, 1964.

Assigned to CVW-21, VAH-10 Det 19 made a deployment to the Gulf of Tonkin aboard the USS "Hancock" (CVA-19) from August 2, 1969 until April 15, 1970. One of its KA-3Bs, BuNo 138974, is seen here with the modex 015 prior to application of the 613 modex used during that cruise.

HEAVY ATTACK SQUADRON TEN
VAH-10 "VIKINGS"

Commissioned at NAS Whidbey Island, Washington, on May 1, 1961, VAH-10 was the last HATRON to be activated. It received its first

A3D-2s six weeks after being organized. As part of CVG-8, the squadron took part in the shakedown cruise of the USS *Constellation* (CVA-64) in March-April 1962. During the following two years, VAH-10 deployed twice to WestPac aboard this carrier. It was there that VAH-10 and the other CVW-8 squadrons provided some of the first naval aircraft to go into combat in Southeast Asia as the *Constellation* was involved in the August, 1964, Gulf of Tonkin Incident.

Back at Whidbey Island in February, 1965, the squadron enjoyed a short leave before departing for the Med aboard the USS *Franklin D. Roosevelt* (CVA-42). Thereafter, VAH-10 became a detachment squadron, with its *Skywarriors* deploying seven times to the Gulf of Tonkin between June, 1966, and August, 1970. Other VAH-10 detachments continued to operate A-3Bs and KA-3Bs aboard carriers of the Sixth Fleet in the Atlantic Ocean and Mediterranean Sea.

The last operational HATRON, as VAH-2 and VAH-4 had been redesignated TACELRONS in November, 1968, VAH-10 finally followed this path and was redesignated VAQ-129 on September 1, 1970.

Bearing the nickname "Luck of the Irish" on its nose, the same KA-3B of VAH-10 Det 19 as illustrated at the bottom of the previous page, is shown here as it appeared on April 15, 1970, on the day of its return from its deployment to the Gulf of Tonkin.

HEAVY ATTACK SQUADRON ELEVEN
VAH-11 "CHECKERTAILS"

HATRON Eleven was formed at NAS Sanford, Florida, on November 1, 1955 from Composite Squadron Eight and initially flew North American AJ-1s. The unit received its first A3Ds in November, 1957.

Remaining homeported at NAS Sanford throughout its existence, VAH-11 deployed to the Med five times aboard the USS *Franklin D. Roosevelt* (CVA-42) and once each aboard the USS *Independence* (CVA-62) and the USS *Forrestal* (CVA-59). Between August, 1962, and January, 1965, the squadron was divided into two units: one with six *Skywarriors* performing all the normal duties of a Heavy Attack Squadron and the other taking up an operational readiness posture from other HATRONS while they converted from A-3s to A-5s. In turn, VAH-11 transitioned to the RA-5C and was redesignated RVAH-11 in July, 1966.

Equipped with "Skywarriors" for just short of nine years, the "Checkertails" of VAH-11 made eight deployments to the Med with their A3D-2s/A-3Bs. One of their A3D-2s is seen here at an air show at NAS Floyd Bennett, New York, on May 19, 1962, when the squadron was shore-based at NAS Sanford between its 3rd and 4th cruise.

A tanker-configured A-3B from VAH-11 is shown taking part in a refueling exercise over the Caribbean on March 2, 1964, while the squadron was working up prior to its fifth and last deployment aboard the USS "Franklin D. Roosevelt" (CVA-42). The tail code AB identifies the aircraft as assigned to CVW-1.

First deploying aboard the USS "Kitty Hawk" (CVA-63) during this carrier's shakedown cruise, VAH-13 made two operational cruises aboard CVA-63 prior to converting to RA-5Cs and being redesignated RVAH-13 in November, 1964. One of its A-3Bs is seen here landing at NAS Whidbey Island.

HEAVY ATTACK SQUADRON THIRTEEN
VAH-13 "BATS"

On January 3, 1961, VAH-13 was commissioned at NAS Sanford, Florida, to fly A3D-2s. The squadron was assigned to CVG-11 upon completing its initial training and deployed on the USS *Kitty Hawk* (CVA-63) for her shakedown cruise in the Caribbean. Shortly thereafter VAH-13 transferred to the Pacific Fleet with the *Kitty Hawk* and its homeport became NAS Whidbey Island, Washington.

For the next three years, the "Bats" maintained readiness requirements and served with the Seventh Fleet. However, in anticipation of its transition to the RA-5C, VAH-13 moved back to NAS Sanford in August, 1964. The squadron was redesignated RVAH-13 on November 1, 1964.

This A-3B of VAH-13 was photographed on December 20, 1963, when the squadron was deployed to WestPac aboard the USS "Kitty Hawk" (CVA-63). The squadron and its air wing, CVW-11, returned to CONUS on June 16, 1964. Two months later VAH-13 moved to NAS Sanford to transition to the RA-5C.

With one of the dirigible hangars at NAS Moffett Field serving as a backdrop for this May, 1962, photograph, this conspicuously marked A3D-1 is seen while it was used for crew training by VAH-123. Less than six years later, by which time it was designated A-3A, this aircraft was salvaged at NARF Alameda.

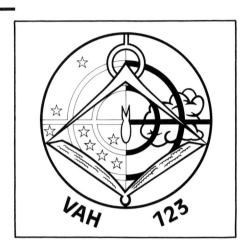

Photographed during a stopover at NAS Adak in the Aleutians on September 8, 1966, this TA-3B was then used as a crew trainer by VAH-123. Twenty years later, BuNo 144859 was still fulfilling this role while operated by VAQ-33 from NAS Key West. The aircraft is seen here in its original configuration with wing pylons for T-65 training shapes or Aero 8A practice bombs.

HEAVY ATTACK SQUADRON ONE TWO THREE
VAH-123 "PROS"

The transition of HATRONs from the propeller-driven *Savage* to the jet-powered *Skywarrior* created the need for jet transitional training. To accomplish this training, HATUPAC (Heavy Attack Training Unit, Pacific) was commissioned on June 15, 1957, at NAS North Island, California; two weeks later HATUPAC moved to NAS Whidbey Island, Washington. The unit was redesignated VAH-123 on June 29, 1959.

Operating as a part of Replacement Carrier Air Group Twelve throughout its existence, VAH-123 retained as its primary mission the training of replacement pilots, bombardier/navigators, crewman/navigators, and maintenance personnel for all Pacific Fleet units using *Skywarriors*. To perform this mission the Pros initially flew Douglas F3D-2Ts, Grumman F9F-8Ts, and Lockheed P2V-3Bs to complement its A3D-1s. Soon, however, the squadron primarily operated various *Skywarrior* models (mainly A-3As, A-3Bs, and TA-3Bs) and the other types were phased out.

Over the years, the activities of VAH-123 were broadened, with low-level navigation and "loft" bombing being added to its syllabus in 1961, and aerial refueling gaining in importance during the mid-sixties. Moreover, in 1964, VAH-123 gained sole responsibility for A-3 training as VAH-3 converted to *Vigilantes*. For 13 months beginning in August, 1966, VAH-123 was also responsible for A-6 crew training, but this duty was later transferred to VA-128.

VAH-123 was decommissioned on February 1, 1971 and its A-3 training responsibility was transferred to VAQ-130.

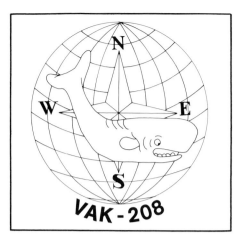

AERIAL REFUELING SQUADRON TWO ZERO EIGHT
VAK-208 "JOCKEYS"

When the Naval Air Reserve was reorganized in 1970 Tactical Electronic Warfare Squadron 208 was commissioned on July 31 at NAS Alameda, California, as part of Carrier Air Wing Twenty. Equipped with KA-3Bs to support CVWR-20 operations, VAQ-208 also began providing air refueling and pathfinding for Navy and Marine tactical aircraft being ferried to Southeast Asia. During the 1971 India-Pakistan conflict and the 1973 Yom Kippur War, VAQ-208 jointly with VAQ-308 provided crews and aircraft to support emergency deployments of U.S. naval aircraft. During the seventies, the squadron also provided tanking platforms for test and development of the Lockheed S-3A *Viking* and Grumman F-14A *Tomcat*. On October 1, 1979, in recognition of its primary use as a refueling and pathfinding unit, the squadron was redesignated VAK-208. Still assigned to CVWR-20, it operates worldwide from NAS Alameda.

Seen here at Offutt AFB, Nebraska, BuNo 147657 successively was an A3D-2, an A-3B, a KA-3B, and an EKA-3B prior to being modified back to the KA-3B configuration. The unit's marking on the rear fuselage identifies the aircraft as belonging to VAK-208, but the insignia below the cockpit still reads VAQ-208.

Photographed on the ramp of the Missouri ANG in St. Louis, BuNo 147665 also went through several configuration changes prior to appearing once again as a KA-3B in February, 1976. The forward fuselage panels above the ATM exhausts show signs of the ECM blisters which were fitted when the aircraft was an EKA-3B.

AERIAL REFUELING SQUADRON THREE ZERO EIGHT
VAK-308 "GRIFFINS"

VAQ-308 was commissioned at NAS Alameda, California, on May 2, 1970, as a sister unit of VAQ-208 and was assigned KA-3Bs for similar duties with CVWR-30. The AERREFRON designation was adopted on October 1, 1979, thus making VAK-208 and VAK-308 the only Navy units ever to have been given air refueling as their primary mission. The squadron which has supported peacetime operations as well as critical emergency deployments (e.g., during the India-Pakistan conflict in 1971 and the transfer of aircraft to Israel during the Yom Kippur War in 1973), is still on the CVWR-30 roster.

Still evidencing signs of its earlier life as an EKA-3B, this KA-3B from VAQ-308 was photographed on the "Whale Country" ramp at NAS Alameda. BuNo 142659 was lost in a take-off accident at Alameda on July 10, 1977. Note the attractive presentation of the November Delta tail code of CVWR-30.

BuNo 147660, in which the author had his first "Skywarrior" ride in November, 1981, taxies out in October, 1982, when the A-3 community was assembled at NAS Alameda to celebrate the "Whale's" 30th anniversary. Ten months later, while being overhauled at NARF Alameda, this aircraft was struck off due to excessive corrosion.

For night operations over hostile territory, several RA-3Bs from VAP-61 were given a black finish as illustrated by BuNo 144840. It was photographed in a partially completed revetment at Da Nang AB in 1968. For ground personnel, the "Whale's" high wing provides some welcome relief from the Vietnamese heat.

Seen here on the ramp at NAS Agana, Guam, on September 26, 1961, this A3D-2P from VAP-61 was lost at sea on October 14, 1967, after having been hit by North Vietnamese 37mm anti-aircraft guns during a day photographic sortie over the coastal region in Route Package IV. This was the third of four RA-3Bs lost in combat by VAP-61.

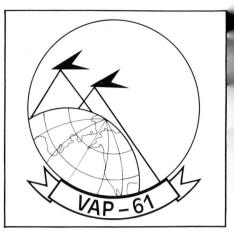

HEAVY PHOTOGRAPHIC SQUADRON SIXTY-ONE
VAP-61 "WORLD FAMOUS"

Tracing its ancestry to Photographic Squadron Five (VD-5), which had been commissioned in June, 1944, and to Composite Squadron 61 (VC-61), which had been organized in January, 1949, this unit was first designated Heavy Photographic Squadron 61 (VAP-61) on July 2, 1956. Homeported at NAS Agana, Guam, VAP-61 was redesignated Composite Photographic Squadron 61 (VCP-61) on July 1, 1959 in preparation for its transition from the AJ-2P to a mix of F8U-1Ps and A3D-2Ps. Two years later, the *Crusaders* were transferred out and the squadron was once again designated VAP-61.

During the Southeast Asia War, VAP-61 sent RA-3B detachments to DaNang AB, South Vietnam, and aboard CTF-77 carriers for operations over North Vietnam and the Ho Chi Minh Trail. Other operational detachments operated at one time or another from locations throughout the Pacific as well as on the Asian mainland, CONUS and Bermuda. VAP-61 was decommissioned on July 1, 1971.

Banking away from the photographer, this RA-3B from VAP-62 shows to advantage the numerous camera ports on the lower sides and beneath its forward fuselage. Close inspection reveals the doors of the aft bomb bay in which a variety of flash bombs and flash cartridges for night photography were carried.

HEAVY PHOTOGRAPHIC SQUADRON SIXTY-TWO
VAP-62 "TIGERS"

Commissioned as Photographic Squadron 62 (VJ-62) on April 10, 1952, this squadron was redesignated VAP-62 on July 2, 1956 while

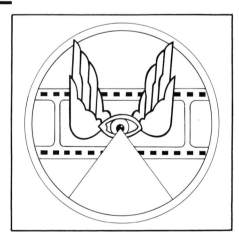

homeported at NAS Norfolk, Virginia. It moved to NAS Jacksonville, Florida, in August, 1957 and soon thereafter evaluated the YA3D-1P prototype.

After transitioning from the AJ-2P in late 1959, VAP-62 retained its A3D-2P/RA-3Bs until its inactivation on October 15, 1969. During this ten-year period, in addition to performing its normal duties from shore bases and carriers, the squadron undertook several special assignments (e.g., photographing the launch of early manned satellites, cartographic survey of the Great Lakes region, photographic survey of flood damage, and tracking and seeding of hurricanes). Moreover, beginning in the fall of 1966, VAP-62 sent RA-3B detachments to Southeast Asia to operate under the control of VAP-61.

Prior to being given to VAQ-33, the "electronic aggressor" mission was undertaken for the "Fleet Electronic Warfare Support Group" (FEWSG) by McDonnell Douglas which specially modified two A-3Bs and flew them under contract. One of these aircraft, BuNo 138922, is seen here taxiing at NAS Alameda on April 23, 1970.

TACTICAL ELECTRONIC WARFARE SQUADRON THIRTY-THREE
VAQ-33 "FIREBIRDS"

Providing realistic electronic warfare simulation during fleet exercises, VAQ-33 traces its origin to the commissioning of Composite Squadron Thirty-Three in May, 1949. The unit, which initially flew Grummann TBM-3Es and later became a major user of multi-seat Douglas *Skyraiders*, was redesignated VA(AW)-33 in July, 1956, VAW-33 in June, 1959, and VAQ-33 in February, 1968.

Making the last *Skyraider* carrier deployment in 1969 aboard the USS *John F. Kennedy*, VAQ-33 was reorganized at NAS Norfolk in early 1970 to become an "electronic aggressor" as the flying unit of the *Fleet Electronic Warfare Support Group* (FEWSG). To fulfill its new missions VAQ-33 has since operated several types of specially modified aircraft including A-3Bs (first assigned in 1970), ERA-3Bs (the unit's core equipment), KA-3Bs, and TA-3Bs. In October, 1977, VAQ-33 also assumed responsibility as the A-3 Fleet Replacement Squadron (FRS) and Fleet Replacement Aviation Maintenance Program (FRAMP). In 1978, most of the unit's activities were transferred to NAS Oceana, Virginia, and in 1980, VAQ-33 was finally relocated at NAS Key West, Florida.

TA-3B, BuNo 144856, of VAQ-33. Basic ground support equipment in the form of cooling blowers is visible in the foreground. Unit insigne is visible on the vertical fin. Aircraft colors are standard, including light gray upper surfaces and white under surfaces.

When VAQ-34 was established on March 1, 1983, the ERA-3Bs which were partially to equip this electronic warfare training unit were not yet ready. Hence, its initial equipment was comprised of a variety of aircraft, including this KA-3B photographed at NAS Point Mugu on October 23, 1983.

TACTICAL ELECTRONIC WARFARE SQUADRON THIRTY-FOUR
VAQ-34 "ELECTRIC HORSEMEN"

On March 1, 1983, as the need for electronic warfare training increased, VAQ-34 was established at NAS Point Mugu to support West Coast operations. The latest *Skywarrior* squadron flew KA-3Bs prior to receiving its four ERA-3Bs. It is also equipped with six Vought A-7Ls and one KA-3B. Like those of VAQ-33, VAQ-34s aircraft bear the GD tail code of their common operational commander, FEWSG.

Equipped with one of the more recent chaff dispensing tail configurations and several other EW-related modifications, ERA-3B, BuNo 144838, of VAQ-34, is seen during a training mission off the coast of Point Mugu, California. Flying "Flashback-213" is LCdr. Rob Ingalls with LCdr. Jerry Noble, and ADCS Bob Caouette.

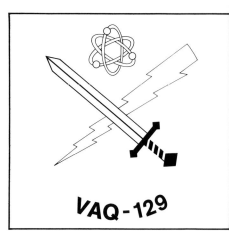

Bearing markings as applied earlier to VAH-10's aircraft, this EKA-3B of VAQ-129 Det 19 assigned to CVW-21 aboard the USS "Hancock" (CVA-19) was photographed over the Gulf of Tonkin on March 18, 1971. Stored at MASDC for three years in 1982-85, BuNo 142650 was returned to service in April, 1985.

TACTICAL ELECTRONIC WARFARE SQUADRON ONE TWENTY-NINE
VAQ-129 "VIKINGS"

When VAH-10, the last *Skywarrior* HATRON, was redesignated VAQ-129 on September 1, 1970, at NAS Whidbey Island, Washington, its Det 38 was aboard the USS *Shangri La* on Yankee Station. Thus, on its commissioning day the new squadron already had a detachment on the line as the personnel and KA-3Bs of Det 38 remained aboard the *Shangri La* for another 15 weeks following their paper transfer to VAQ-129. While preparing for the arrival of its EA-6Bs, the new squadron continued to operate KA/EKA-3Bs for a few months and sent *Skywarrior* detachments aboard the USS *Hancock* (CVA-19) for operations in the Gulf of Tonkin and aboard the USS *Saratoga* (CVA-60) for a Med cruise. Meanwhile, the first EA-6B was received at Whidbey Island in January, 1971, and thereafter VAQ-129 became the *Prowler* Fleet Replacement Squadron (FRS).

Bearing the BuNo 142634, this EKA-3B from VAQ-130 Det 4 was photographed on the port catapult of the USS "Enterprise" (CVAN-65) in June, 1971, at the start of an eight-month combat cruise. It was lost on January 21, 1973, in a catapult accident aboard the USS "Ranger" (CVA-61) while operated by VAQ-130 Det 4 in the Gulf of Tonkin.

In addition to deployments aboard six carriers operating in the Gulf of Tonkin and the South China Sea, EKA-3Bs from VAW-13 Dets flew combat support missions from Da Nang AB where BuNo 147667 was photographed. The squadron was redesignated VAQ-130 on October 1, 1968, shortly after this photograph was taken.

TACTICAL ELECTRONIC WARFARE SQUADRON ONE THIRTY
VAQ-130 "ZAPPERS"

Commissioned at NAS Agana, Guam, on September 1, 1959 and transferred to NAS Alameda, California, in July, 1961, Carrier Airborne Early Warning Squadron Thirteen flew a variety of aircraft during its first eight years. In the spring of 1967 this unit received the first EKA-3Bs and KA-3Bs and soon began to deploy its *Skywarriors* to provide aerial refueling and electronic countermeasures while continuing to operate Douglas EA-1Fs until November, 1970. From November, 1967, until March, 1969, six VAW-13 detachments operated EKA-3Bs aboard the USS *America* (CVA-66), *Bon Homme Richard* (CVA-31), *Constellation* (CVA-64), *Enterprise* (CVAN-65), *Hancock* (CVA-19), and *Ranger* (CVA-61).

On October 1, 1968, the squadron was redesignated VAQ-130 and 14 of its detachments served aboard carriers in the Gulf of Tonkin until the return of Det 4 in June, 1973. From May to October, 1974, VAQ-130 Det 4 made the final EKA-3B peacetime deployment aboard the USS *Ranger* (CVA-61). Following the decommissioning of VAH-123 in February, 1971, VAQ-130 also took over the responsibility for training A-3 crews. VAQ-130 stood down on June 30, 1974 and, moving to NAS Whidbey, Washington, began its transition to the EA-6B.

CARRIER AIRBORNE EARLY WARNING SQUADRON THIRTEEN
VAW-13 "ZAPPERS"
See history of VAQ-130.

Although assigned to the Atlantic Fleet, the USS "America" (CVA-66) made three war cruises, two with LANT Air Wings (CVW-6 and CVW-8) and one with a PAC Air Wing (CVW-9). This EKA-3B of VAW-13 Det 66 bears the tail code of CVW-6 with which the USS "America" deployed from April 10, to December 16, 1968.

TACTICAL ELECTRONIC WARFARE SQUADRON ONE THIRTY-ONE
VAQ-131 "LANCERS"

VAQ-131 was organized at NAS Alameda, California, on November 1, 1968, when VAH-4 was redesignated. Initially equipped with KA/EKA-3Bs, the new TACELRON sent one *Skywarrior* detachment to the Gulf of Tonkin aboard the USS *Kitty Hawk* (CVA-63) and one detachment to the Med aboard the USS *J. F. Kennedy* (CVA-67). In May, 1971, VAQ-131 moved to NAS Whidbey Island, Washington, to transition to the *Prowler*.

The six stripes painted around the rear fuselage of this EKA-3B from VAQ-131 were applied to indicate that it was tanker #6 in a multi-tanker Alpha strike. BuNo 142632 was photographed at NAS Alameda in September, 1969, after this TACELRON's return from a war cruise aboard the USS "Kitty Hawk" (CVA-63).

TACTICAL ELECTRONIC WARFARE SQUADRON ONE THIRTY-TWO
VAQ-132 "SCORPIONS"

VAQ-132 came into existence on November 1, 1968, when VAH-2 was redesignated. Homeport for the new TACELRON was NAS Alameda, California, but already the squadron had a detachment on Yankee Station as it inherited Det 64 aboard the USS *Constellation* (CVA-64). While equipped with KA/EKA-3Bs VAQ-132 made two more combat cruises, one each aboard the USS *Enterprise* (CVAN-65) and the USS *America* (CVA-66). For the latter, two of the squadron's KA-3Bs went aboard the carrier in Norfolk, Virginia, for the ocean voyage around South Africa, while three EKA-3Bs transpac'd and rejoined the carrier at Subic Bay in the Philippines. Staying aboard the carrier after seven months of combat operations, VAQ-132 returned to Alameda by way of the Indian Ocean, the South Atlantic, and the carrier's homeport in Norfolk. On January 15, 1971, VAQ-132 moved to NAS Whidbey Island, Washington, and soon thereafter transitioned to the EA-6B.

Departing Alameda on January 6, 1969, for a deployment to the Gulf of Tonkin, the USS "Enterprise" suffered a major fire during ORI off Hawaii on January 14. After repairs, CVAN-65 got on the line at the end of March. One of the EKA-3Bs from VAQ-132 which took part in this ill-fated cruise is seen recovering aboard the "Enterprise".

Photographed at NAS Alameda on October 23, 1970, this EKA-3B is one of the aircraft from VAQ-133 which embarked two weeks later aboard the USS "Kitty Hawk" (CVA-63) for an eight-month war cruise to the Gulf of Tonkin. BuNo 147655 bears the NH tail code of CVW-11.

As part of CVW-14, the "Wizards" of VAQ-133 deployed aboard the USS "Constellation" (CVA-64) from August 11, 1969 until May 8, 1970. Their BuNo 147663 was photographed at NAS Barbers Point while "Connie" was transiting in Hawaiian waters on its way to war.

TACTICAL ELECTRONIC WARFARE SQUADRON ONE THIRTY-THREE VAQ-133 "WIZARDS"

Commissioned at NAS Alameda, California, on March 4, 1969, VAQ-133 was the fourth squadron created for the primary mission of detecting and jamming hostile radar signals. For the next 29 months, VAQ-133 flew KA/EKA-3Bs and during that period deployed twice to the Gulf of Tonkin, once aboard the USS *Constellation* (CVA-64) and once aboard the USS *Kitty Hawk* (CVA-63). Beginning in August, 1971, the squadron stood down for one year. On August 4, 1972, it was reactivated at NAS Whidbey Island, Washington, as a *Prowler* squadron.

This pair of EKA-3Bs from VAQ-134 was photographed in June, 1971, at NAS Alameda after the "Garudas" had returned from the Gulf of Tonkin where they had deployed aboard the USS "Ranger" (CVA-61) as part of CVW-2. The squadron stood down one month later in preparation for its move to NAS Whidbey Island and its transition to EA-6Bs.

EKA-3B, BuNo 142652, with hook down attempts a landing aboard the USS "Ranger" (CVA-61) in the Gulf of Tonkin sometime during 1970. Noteworthy are the extended speedbrakes, tail wheel, and flaps. From the angle of this photo it appears that the aircraft was too high for its tail hook to engage a cross-deck pendant. Visible on the carrier deck are the trapping cables. Aircraft weight dictates the playout speeds for the cables.

TACTICAL ELECTRONIC WARFARE SQUADRON ONE THIRTY-FOUR VAQ-134 "GARUDAS"

Established on June 17, 1969, at NAS Alameda, California, VAQ-134 flew KA-3Bs and EKA-3Bs for 25 months. During this period it made two combat cruises aboard the USS *Ranger* (CVA-61). The Garudas stood down in July, 1971, and moved to NAS Whidbey Island, Washington, in May, 1972, to fly EA-6Bs.

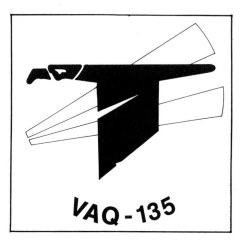

TACTICAL ELECTRONIC WARFARE SQUADRON ONE THIRTY-FIVE
VAQ-135 "BLACK RAVENS"

On May 15, 1969, one of VAQ-130 detachments was enlarged into a new TACELRON, VAQ-135, homeported at NAS Alameda, California. The new unit first deployed to Southeast Asia aboard the USS *Coral Sea* (CVA-43) in September, 1969. In July, 1970, as other TACELRONs were transitioning to the EA-6B, VAQ-135 was reorganized into a permanently shore-based component and five sea-going detachments. From November, 1971, until January, 1974, these detachments deployed to Yankee Station twice aboard the USS *Coral Sea* (CVA-43) and USS *Hancock* (CVA-19), and once aboard the USS *Kitty Hawk* (CVA-63). VAQ-135 detachments also deployed to the Med aboard the USS *F. D. Roosevelt* (CVA-42), USS *Forrestal* (CVA-59), USS *America* (CVA-66), and USS *J. F. Kennedy* (CVA-67). Moving to NAS Whidbey Island, Washington, in September, 1973, VAQ-135 received its first EA-6Bs in July, 1974.

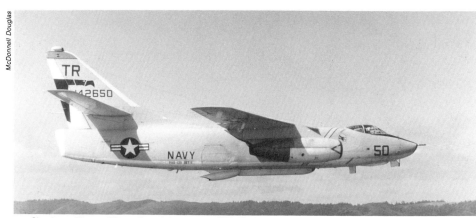

Showing evidence of its earlier existence as an EKA-3B, this KA-3B from VAQ-135 Det 1 bears the tail code TR used by the squadron when shore-based at NAS Alameda. When embarked aboard the USS "Franklin D. Roosevelt" (CVA-42) as part of CVW-6, VAQ-135 Det 1 used the tail code AE.

This EKA-3B from VAQ-135 Det 5 was photographed on take-off at NAF Atsugi, Japan, on July 19, 1973. At that time this detachment was assigned to CVW-21 aboard the USS "Hancock" (CVA-19). Note the undercarriage's retraction geometry.

COMPOSITE PHOTOGRAPHIC SQUADRON SIXTY-THREE
VCP-63 "EYES OF THE FLEET"

Commissioned on January 20, 1949, as Composite Squadron Sixty-One, this unit was successively redesignated Fighter Photographic Squadron 61 in July, 1956, and Composite Photographic Squadron Sixty-Three on July 1, 1959. At that time, VCP-63 received A3D-2Ps to complement its inventory of F8U-1Ps. However, the photographic version of the *Skywarrior* was phased out after only two years with VCP-63, and the squadron's designation was changed to VFP-63 on July 1, 1961.

Photographed above NAS Miramar, VCP-63's shore base, this A3D-2P is one of the "Skywarriors" which partially equipped Composite Photographic Squadron Sixty-Three for two years. VCP-63 lost its A3D-2Ps in July, 1961, after becoming VFP-63.

This camouflaged RA-3B was inherited by VQ-1 from VAP-61 in June, 1961, but was almost immediately sent to the Military Aircraft Storage & Disposition Center (MASDC) at Davis-Monthan AFB. It was retrieved from storage in 1982 to be modified as an ERA-3B for service with VAQ-34.

Photographed at Elmendorf AFB, Alaska, on May 20, 1964, BuNo 146459 is one of the EA-3Bs operated by VQ-1 in the Tactical Airborne Signal Exploitation (TASES) role. Accepted by the Navy in July, 1960, this long-lived aircraft was still in service with VQ-1 more than 26 years later.

FLEET AIR RECONNAISSANCE SQUADRON ONE
VQ-1 "WORLD WATCHERS"

VQ-1 traces its origin to the organization of a special project division at NS Sangley Point in the Philippines during October, 1951. The squadron itself was commissioned at Sangley Point on June 1, 1955, as Electronics Countermeasures Squadron One and, equipped with Martin P4M-1Qs, moved to MCAF Iwakuni, Japan, in October, 1955.

Having flown *Skywarriors* ever since receiving its first A3D-1Qs in November, 1956, VQ-1 shares with VQ-2 the distinction of being the squadron to have been equipped with the same type of aircraft for the longest period. However, the squadron also flew Lockheed WV-2/WC-121/EC-121s from February, 1960, until the summer of 1974, and continues to fly Lockheed EP-3s which it received in March, 1969.

The squadron moved to NAS Atsugi, Japan, in July, 1960, and took its current designation. Four years later, VQ-1 began deploying EA-3Bs in support of combat operations in Southeast Asia, with its aircraft deploying to CTF-77 carriers and to Da Nang AB in South Vietnam to perform in the TASES (Tactical Airborne Signal Exploitation System) role. In addition, after moving to NAS Agana, Guam, in June, 1971, VQ-1 inherited the RA-3Bs and mission of the decommissioned VAP-61. This role, however, was relinquished in the summer of 1974 and the RA-3Bs were sent back to CONUS.

Still providing electronic reconnaissance missions in support of fleet operations, VQ-1 frequently deploys detachments aboard carriers (this was notably the case during operations in the aftermath of the Iranian revolution and following the Soviet invasion of Afghanistan). At the beginning of 1987, VQ-1 was still flying *Skywarriors* (EA-3Bs and TA/VA-3Bs) alongside its Lockheed EP-3s.

VQ-2 EA-3B, BuNo 146453, remains active and is therefore one of the "long-lived Skywarriors" still on inventory. It is seen at NAS Alameda, California, on June 3, 1981, where it had been reworked by the Naval Air Rework Facility in order to increase its service life and update many of its onboard systems. Noteworthy are the no-longer visible BuNo and the distinctive VQ-2 vertical fin patch.

FLEET AIR RECONNAISSANCE SQUADRON TWO
VQ-2 "BATMEN"

Counterpart of the Pacific Fleet's VQ-1, FAIRECONRON TWO has served with the Atlantic Fleet since it was commissioned at Port Lyautey, Morocco, as Electronics Countermeasures Squadron Two on September 1, 1955. A3D-1Qs began supplementing the unit's P4M-1Qs in 1956 and *Skywarriors* have been operated by the squadron ever since. Homeported at NS Rota, Spain, since November, 1958, the squadron was redesignated VQ-2 on January 1, 1960. Over the years, VQ-2 has operated Lockheed EC-121s and EP-3s alongside its *Skywarriors* to provide electronic reconnaissance for the Sixth Fleet.

In addition to its normal operations, including regular deployments since 1965 aboard carriers operating in the Med and in the Atlantic, VQ-2 has undertaken several notable deployments: a deployment to NAS Key West, Florida, in the fall of 1962 to provide SIGINT during the Cuban crisis; the deployment of Det Bravo to Da Nang AB, South Vietnam, for four years beginning in 1965; and deployments to the Eastern Mediterranean and the Indian Ocean during various contingencies (including the confrontation with Libya during April, 1986).

Skywarrior models operated by VQ-2 have included the A3D-1Q, A3D-2Q/EA-3B, and TA-3B.

Photographed on August 8, 1975, on the transient line at NAS North Island, San Diego, California, this EA-3B bears the "Spanish spook" markings of VQ-2, the Fleet Air Reconnaissance Squadron based at NS Rota, Spain. The aircraft is fitted with the larger "Seawing" ventral canoe.

An EA-3B with its ELINT antenna ventral canoe fairing. Distinctive VQ-2 unit patch can be seen adorning the aircraft's vertical fin. Unit provides electronic reconnaissance capability for the Sixth Fleet and has been homeported at NS Rota, Spain since November, 1958.

MISCELLANEOUS UNITS:

FLEET TRANSPORT SQUADRON ONE (VR-1)

At least one transport-configured TA-3B is known to have been operated by this TRANSRON.

AEROSPACE RECOVERY FACILITY (ARF)

The Naval Aerospace Recovery Facility (previously designated National Parachute Test Range, NPTR, as a joint Navy-Air Force unit) at NAF El Centro, California, briefly operated an A-3B *Skywarrior* in support of its activities. Along with its A-3B, NPTR also operated the *Skywarrior* derivative, the Air Force B-66.

COMMANDER, FLEET LOGISTICS SUPPORT WING (COMFLELOGSUPPWING DET WASHINGTON DC)

Beginning in June, 1959, a single VA-3B (BuNo 142672) was operated from NAS Patuxent River as a VIP transport for the CNO, as well as for sundry projects at the Naval Air Test Center. In September, 1965, this aircraft and some personnel were transferred to NAF Washington D.C., thus marking the debut of the present COMFLELOGSUPPWING DET (or CFLSW Det). This detachment was officially formed in July, 1978, when it was attached to the Commander, Reserve Tactical Support Wing (COMRESTACSUPPWING) and received its current designation in May, 1982. Its mission ranges from executive transport of dignitaries—including cabinet members, the CNO, and foreign VIP visitors—to support for medical recovery teams from the Bethesda Naval Hospital, the National Institute of Health, and the Walter Reed Army Hospital. At the end of 1985 this Det. operated two VIP-configured TA-3Bs and two North American CT-39Gs.

FLEET ELECTRONIC WARFARE SUPPORT GROUP (FEWSG)

See VAQ-33 and VAQ-34.

NATIONAL PARACHUTE TEST RANGE (NPTR)

See Aerospace Rescue Facility.

NAVAL AIR DEVELOPMENT CENTER (NADC)

Established on August 1, 1948, at Johnsville, Pennsylvania, the Naval Air Development Center has had the missions of performing research and development in the field of aviation medicine and of developing aircraft electronic, pilotless aircraft, and aviation armament. NADC is known to have operated at least two RA-3Bs.

NAVAL AIRBORNE PROJECT OPERATIONS GROUP (NAPOG)

To provide operational control and airborne support to the Pacific Range Electro-magnetic Signature Studies (Project PRESS), in 1963 NAPOG received an NRA-3B which had been modified by Douglas to mount infrared, visual and ultraviolet sensors in a dorsal turret. Based at Hickam AFB, Hawaii, the aircraft operated from the Kwajalein Army Test Site and the Eniwetok Air Force Auxiliary Airfield.

NAVAL AIR RESERVE UNIT ALAMEDA (NARU ALAMEDA)

In June, 1974, when VAQ-130 moved to NAS Whidbey Island, Washington, to become an EA-6B unit, responsibility for training *Skywarrior* aircrew and maintenance personnel was transferred to the Naval Air Reserve Unit at NAS Alameda, California. Primarily operating TA-3Bs and RA-3Bs, NARU Alameda performed as the A-3 Fleet Replacement Squadron until October, 1977, when the *Skywarrior's* longer than expected life prompted the return of FRS/FRAMP functions to an active unit, VAQ-33.

NAVAL AIR SPECIAL WEAPONS FACILITY/ NAVAL WEAPONS EVALUATION FACILITY (NASWF/NWEF)

Established at Kirtland AFB, New Mexico, in August, 1952, NASWF provided naval participation in various programs involved in the application of nuclear weapons to aircraft. Later redesignated Naval Weapons Evaluation Facility, this facility conducted the special weapons phase of the A3D-1 BIS trials in 1956-57. Subsequently, additional tests were conducted to clear the *Skywarrior* for carriage and release of a variety of nuclear devices.

Accepted by the Navy in October, 1955, BuNo 135413 was initially assigned to the Naval Air Special Weapons Facility (NASWF) at Kirtland AFB, New Mexico, for use in the A3D-1 Service Acceptance Trials, Special Weapons Phase. At the conclusion of these tests, the A3D-1 was cleared for delivering Marks 5, 6, 7, 8, 12, 15 Mod 0, 18, and 91 special weapons. This A3D-1 was lost in an accident while undergoing an acceptance flight after overhaul at NARF Alameda in 1963.

NAVAL AIR TEST CENTER (NATC)

NATESTCEN active participation in the *Skywarrior* program began on November 5, 1954 when one of the XA3D-1s was delivered to NAS Patuxent River, Maryland, for accelerated service tests. Subsequently, NATC tested all *Skywarrior* versions. The Center last used a KA-3B as a tanker in support of the F-18 program.

BuNo 135411 is shown here aboard the USS "Forrestal" (CVA-59) in April, 1956, during the A3D-1 initial carrier suitability trials. The cockpit's hatch is open, as is normal procedure during carrier operations, to facilitate evacuation in the event of ditching.

Bearing the stylized W tail markings applied to aircraft used for weapons trials, this A-3B was still used by the NATC in June, 1970, when this photograph was taken. BuNo 142246 is preserved at the Bradley Air Museum in Connecticut.

BuNo 138905, the fourth A3D-2 was used by the Naval Air Test Center for combined Stability and Control and Aircraft and Engine Performance Trials from November, 1957, until April, 1958.

Accepted in August, 1956, by which time A3D-1s were already in squadron service, BuNo 135431 had a short life for a "Skywarrior" as it was salvaged at NARF Alameda less than 12 years later.

NAVAL AIR TEST FACILITY (NATF)

Established at NAS Lakehurst, New Jersey, on October 1, 1957, NATF was responsible for evaluating and supporting the development of aircraft launching and recovery systems. Among the aircraft types operated in support of its role, NATF was assigned at least one A-3A. In addition, time-expired or otherwise no longer airworthy *Skywarriors* were used at NAS Lakehurst for ground training and non-flying tests.

With its "eight-ball" markings and liberal use of red paint, this A3D-1 (BuNo 135407) from the Naval Air Test Facility at NAS Lakehurst was one of the most colorful "Skywarriors". Built as the first aircraft of the second A3D-1 production contract, it was accepted by the Navy in January, 1955, and was stored at MASDC from July, 1971, until April, 1977, when it was bought for scraps by Kolar Inc.

NAVAL MISSILE CENTER/ PACIFIC MISSILE TEST CENTER (NMC/PMTC)

During 1959, the U.S. Naval Missile Center at NAS Point Mugu, California, received its first two A3D-1s. Since then PACMISTESTCEN has operated highly modified *Skywarriors* (NA-3A, NA-3B, NRA-3B, and NTA-3B) to support several of its activities, including captive flight testing of missile weapon systems, test and evaluation of electronic warfare components and systems, and electronic warfare exercises for the Fleet.

Best known NRA-3B of the Pacific Missile Test Center, bulbous-nosed, BuNo 144825, wings folded, is seen at NAS Point Mugu on April 5, 1982, mounting a twin-pack chaff dispenser system under its left wing. Twin-pack mounting was used irregularly in the A-3 fleet. The ability to twin-pack was usually dictated by the weight and c.g. of the pod.

NAVAL ORDNANCE TEST STATION/ NAVAL WEAPONS CENTER (NOTS/NWC)

In support of its RDT&E activities, NOTS China Lake, California, (later redesignated NWC China Lake) has operated a small number of *Skywarriors* as testbeds for weapons and armament systems. In early 1987 a KA-3B was still in use at China Lake.

BuNo 142404 of the Naval Ordnance Test Station was photographed at NAS Alameda on February 10, 1968. Just over a year later, it was converted to the KA-3B configuration for assignment to a Fleet squadron.

To replace BuNo 142404, NOTS/NWC received BuNo 142630, which is seen here on take-off from NAS Miramar in July, 1980. Designated NA-3B, this aircraft was still with NWC in early 1987.

Over the years, as detailed in Chapter V, other users have included the Grumman Aerospace Corporation, the Hughes Aircraft Company, the Westinghouse Electric Corporation, and the United States Army.

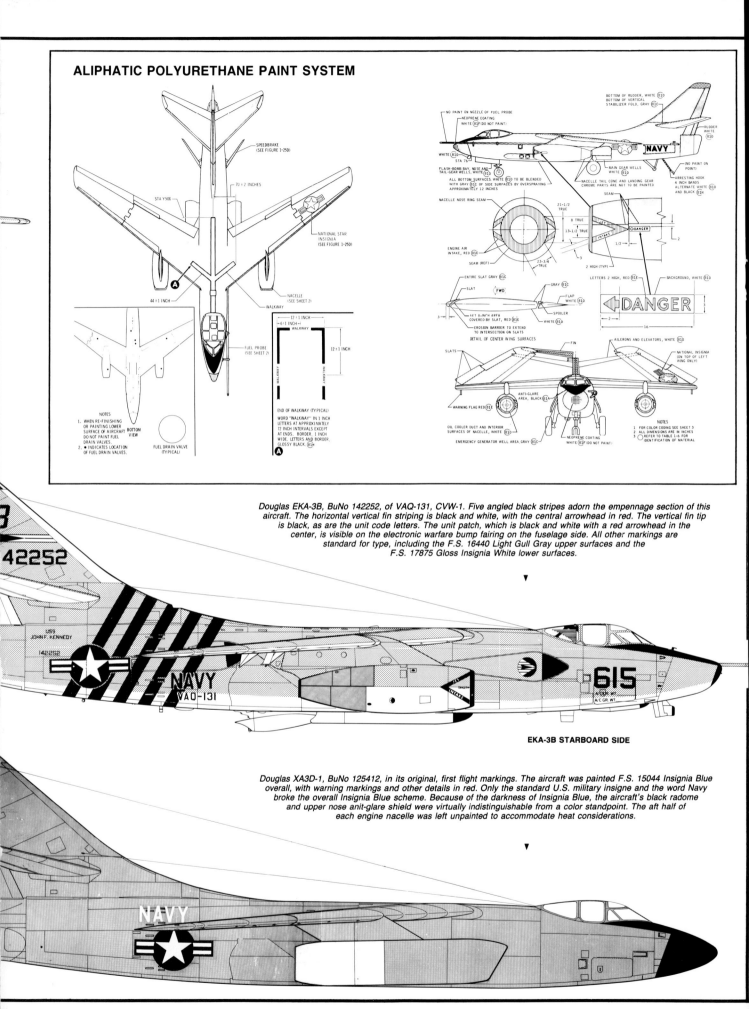

Douglas EKA-3B, BuNo 142252, of VAQ-131, CVW-1. Five angled black stripes adorn the empennage section of this aircraft. The horizontal vertical fin striping is black and white, with the central arrowhead in red. The vertical fin tip is black, as are the unit code letters. The unit patch, which is black and white with a red arrowhead in the center, is visible on the electronic warfare bump fairing on the fuselage side. All other markings are standard for type, including the F.S. 16440 Light Gull Gray upper surfaces and the F.S. 17875 Gloss Insignia White lower surfaces.

EKA-3B STARBOARD SIDE

Douglas XA3D-1, BuNo 125412, in its original, first flight markings. The aircraft was painted F.S. 15044 Insignia Blue overall, with warning markings and other details in red. Only the standard U.S. military insigne and the word Navy broke the overall Insignia Blue scheme. Because of the darkness of Insignia Blue, the aircraft's black radome and upper nose anit-glare shield were virtually indistinguishable from a color standpoint. The aft half of each engine nacelle was left unpainted to accommodate heat considerations.

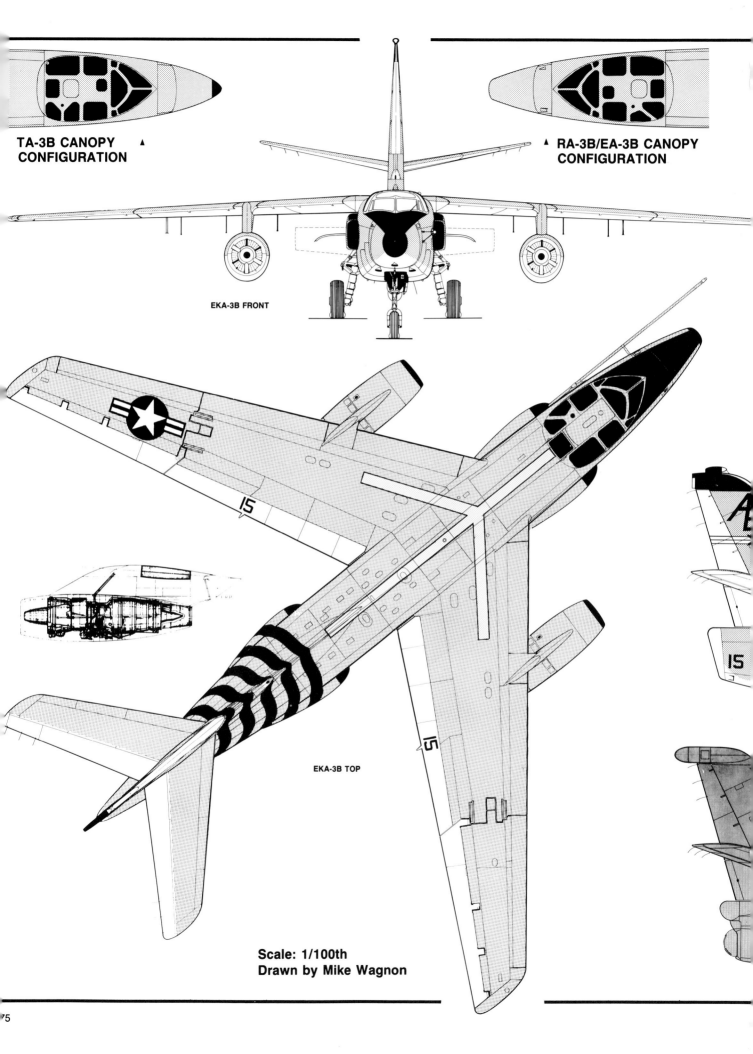

SELECT MARKINGS

Scale: 1/100th
Drawn by Mike Wagnon

Douglas TA-3B, BuNo 144862, of VAQ-130. The aircraft is painted F.S. 16440 Light Gull Gray overall, except for the F.S. 17875 Insignia White nose cap and the unpainted rear segment of each engine nacelle. The anti-glare shield ahead of the windscreen is F.S. 37038 matte Black and the tail code and other related markings are F.S. 17038 Gloss Black/Jet. All other markings are standard for type.

Douglas TA-3B, BuNo 144858, of RVAH-3. The aircraft is painted F.S. 16440 Light Gull Gray on its upper surfaces and F.S. 17875 Insignia White on its lower surfaces. The vertical fin horizontal stripe is F.S. 11136 Insignia Red, the tail code letters are black, and the vertical fin leading edge is Insignia White. Interestingly, the small triangle integral with the Insignia Red engine intake warning lines is in blue, with three overlaid white stars.

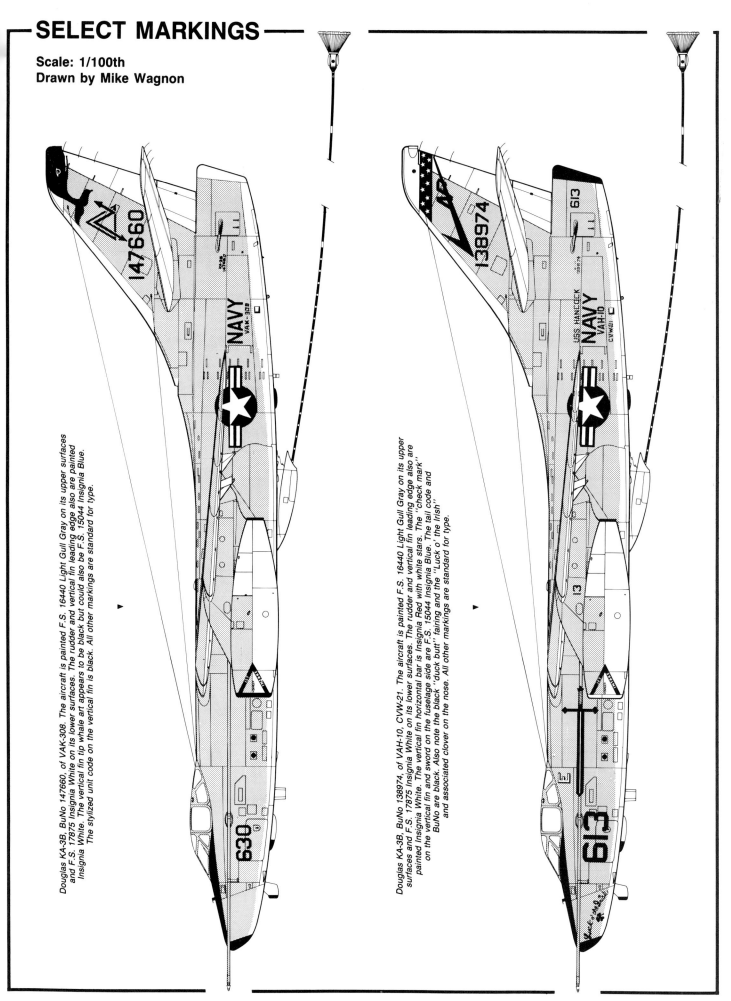

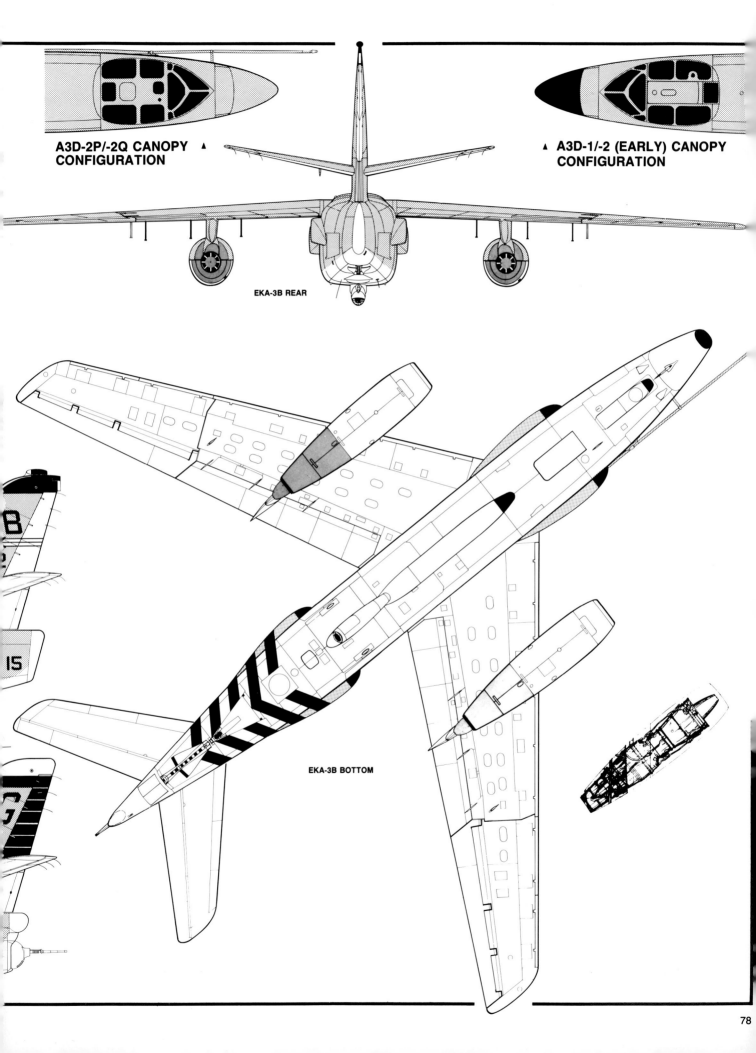

DOUGLAS EKA-3B, BuNo 142252

ALIPHATIC POLYURETHANE PAINT SYSTEM

CODE	ANA	FED-STD 595	DESCRIPTION	FINISH	TABLE 1-6 ITEM NUMBER
		37038	BLACK	CAMOUFLAGE	81A
		36440	LIGHT GULL GRAY	CAMOUFLAGE	81B
		16440	LIGHT GULL GRAY	GLOSSY	81C
		17875	INSIGNIA WHITE	GLOSSY	81D
		11136	INSIGNIA RED	GLOSSY	81E
			WHITE	NEOPRENE ON GLASS FIBER	81F
		15044	INSIGNIA BLUE	GLOSSY	81G
		17038	BLACK	GLOSSY	81H
		12197	INTERNATIONAL ORANGE	GLOSSY	81J
		13538	ORANGE-YELLOW	GLOSSY	81K

WARNING

Make certain maintenance safety precautions are complied with and that no electrical power is connected to aircraft before starting work in cockpit area and on or under aircraft. Failure to comply may result in injury to personnel.

1. PAINTING LARGE AREAS WITH POLYURETHANE FINISH. Prior to painting, areas must be stripped or sanded to remove old paint and surfaces must be pretreated.

   NOTE

   Do not use polyurethane finish on wheels or brakes.

   a. Mask all surfaces that could be damaged by stripping materials, surface treatment, and paint overspray.

   b. Strip old paint from surface to be finished. Refer to NAVAIR 01-1A-509 for procedure.

   c. Remove all corrosion deposits. Refer to table 1-4A and NAVAIR 01-1A-509.

   d. Clean surfaces with water-emulsion cleaning compound (item 18A, table 1-6).

   e. Conduct water-break test (MIL-F-18264) to make certain surfaces are thoroughly clean. Refer to table 1-5A.

   f. Apply chemical surface treatment. Refer to table 1-5A.

   NOTE

   Cleanliness and proper preparation of base surface is of great importance if good adhesion is to be obtained.

   g. Apply wet spray coat of pretreatment coating (item 82, table 1-6) over anodized, titanium, and stainless steel surfaces to be painted. Air dry for 1 hour. Refer to table 1-5A for mixing procedure.

   h. Apply epoxy polyamide primer coating (item 81M, table 1-6) over pretreatment coating. Air-dry minimum of 1 hour prior to application of topcoat. Refer to table 1-5A for mixing ratios of components.

   i. Scuff-sand prime coat with 400-grit aluminum oxide paper to remove any rough areas or overspray. Remove residue with tack cloths or acrylic thinner (item 101, table 1-6).

   NOTE

   Polyurethane paint is applied as an erosion barrier on all metal leading edges of the wing, stabilizers, fins, and engine nacelles. An additional final coat must be applied in addition to the regular paint coats.

   j. Apply coat of polyurethane finish (MIL-C-81773(AS)) over primer. Refer to table 1-5A for mixing instructions. Allow coat to dry 15 to 45 minutes.

   k. Apply first final coat to areas that are subject to erosion to prevent overspray on finish coat. Allow to air-dry 3 to 4 hours.

   l. Apply final full coat of polyurethane finish and allow to air-dry for 6 to 8 hours.

   NOTE

   Insignia white polyurethane has a slight translucent characteristic that may require additional drying time between coats to obtain an opaque finish.

2. TOUCH-UP OR OVERPAINT OF EPOXY POLYAMIDE FINISH WITH ALIPHATIC POLYURETHANE FINISH.

   WARNING

   Make certain maintenance safety precautions are complied with and that no electrical power is connected to aircraft before starting work in cockpit area and on or under aircraft. Failure to comply may result in injury to personnel.

   NOTE

   Aliphatic polyurethane finish will not adhere to acrylic nitrocellulose lacquer finish. Aircraft with lacquer finishes must be stripped prior to applying polyurethane finish.

   a. Mask and seal all areas in vicinity of touchup to prevent paint overspray and seepage of stripping materials.

   b. Clean areas to be finished and remove decals and lacquer markings as necessary.

   c. Remove all corrosion products. (Refer to table 1-4A and NAVAIR 01-1A-509).

1. PAINT FINISH SYSTEMS

   Paint finish system information consists of data required for applying paint, insignias, decals, and stenciling. Paint systems discussed are:

   Acrylic nitrocellulose lacquer system

   Aliphatic polyurethane system

2. ACRYLIC NITROCELLULOSE LACQUER SYSTEM. The acrylic nitrocellulose lacquer finish is used in some interior areas of the aircraft fuselage, wings, cockpit and externally on insignias and markings and exterior surfaces. See figures 1-25C and 1-25D.

   a. Stripping. To strip acrylic nitrocellulose lacquer finishes, refer to NAVAIR 01-1A-509 for proper procedures.

   b. Surface Treatment. Surface treatment must be applied when the original chemical treatment is damaged. Surfaces must be masked or sealed to prevent seepage of chemical solutions into faying edges where damage would result. Alodine 1200 (MIL-C-5541) is used for the surface treatment of aluminum.

   c. Touchup. Touchup of acrylic nitrocellulose lacquer finishes is required for small areas (approximately 4 square inches) that are slightly abraded but with no breaks in the coating. Refer to NAVAIR 01-1A-509 for proper touchup procedure.

   d. Painting Large Areas. Painting large areas of the aircraft with acrylic nitrocellulose lacquer system (MIL-L-19537, gloss and MIL-L-1953*, camouflage) requires surface stripping and surface treatment, if original surface treatment is damaged. Refer to NAVAIR 01-1A-509 for surface treatment and finish application.

3. ALIPHATIC POLYURETHANE FINISH SYSTEM. The aliphatic polyurethane system is a glossy finish used on the exterior of the aircraft. The polyurethane finish is used on all aircraft that are authorized to receive exterior paint finishes at Naval Air Rework Facility, Alameda. The finish system consists of two coats of polyurethane (MIL-C-81773(AS)) applied over one coat of epoxy primer Specification MIL-P-23377. The polyurethane system is compatible with standard touchup procedures used for epoxy polyamide systems.

   NOTE

   Aliphatic polyurethane finish will not adhere to acrylic nitrocellulose lacquer finish. Aircraft with lacquer finishes must be stripped prior to applying polyurethane finish.

   a. Stripping. Stripping of previously painted surfaces may be performed by complying with procedures given in NAVAIR 01-1A-509. Polyurethane finish is compatible with epoxy polyamide provided surface is processed per Specification MIL-P-23377.

   b. Surface Treatment. Surface treatment prior to application of aliphatic polyurethane finish system must be applied on metal surfaces that have the original surface treatment removed or damaged. Surface treat bare metal with chrome conversion coating MIL-C-5541, Type II.

   c. Touchup and Painting Procedures. For touchup and painting large areas, see figure 1-25B.

Scale: 1/100th
Drawn by Mike Wagnon

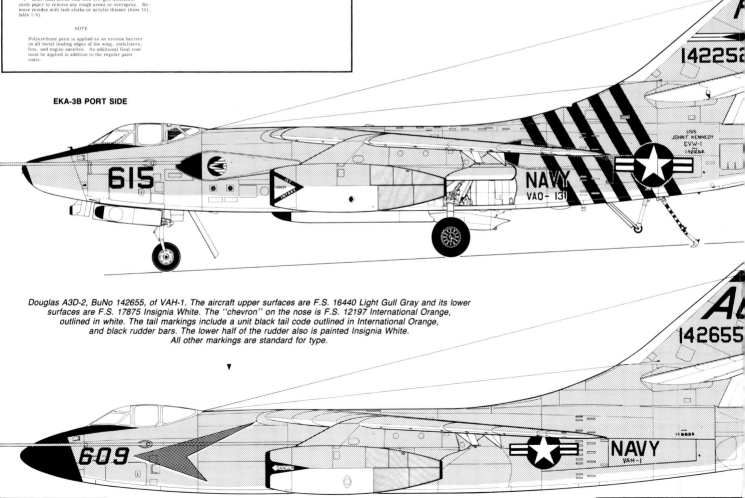

EKA-3B PORT SIDE

Douglas A3D-2, BuNo 142655, of VAH-1. The aircraft upper surfaces are F.S. 16440 Light Gull Gray and its lower surfaces are F.S. 17875 Insignia White. The "chevron" on the nose is F.S. 12197 International Orange, outlined in white. The tail markings include a unit black tail code outlined in International Orange, and black rudder bars. The lower half of the rudder also is painted Insignia White. All other markings are standard for type.

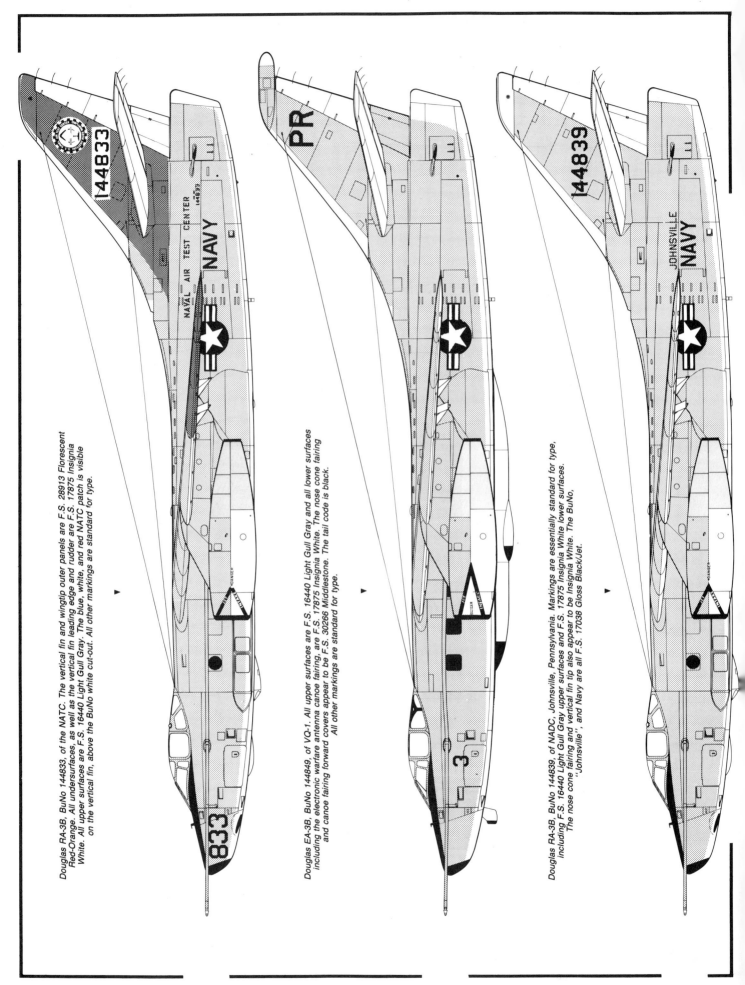

Douglas RA-3B, BuNo 144833, of the NATC. The vertical fin and wingtip outer panels are F.S. 28913 Florescent Red-Orange. All undersurfaces, as well as the vertical fin leading edge and rudder are F.S. 17875 Insignia White. All upper surfaces are F.S. 16440 Light Gull Gray. The blue, white, and red NATC patch is visible on the vertical fin, above the BuNo white cut-out. All other markings are standard for type.

Douglas EA-3B, BuNo 144849, of VQ-1. All upper surfaces are F.S. 16440 Light Gull Gray and all lower surfaces including the electronic warfare antenna canoe fairing, are F.S. 17875 Insignia White. The nose cone fairing and canoe fairing forward covers appear to be F.S. 30266 Middlestone. The tail code is black. All other markings are standard for type.

Douglas RA-3B, BuNo 144839, of NADC, Johnsville, Pennsylvania. Markings are essentially standard for type, including F.S. 16440 Light Gull Gray upper surfaces and F.S. 17875 Insignia White lower surfaces. The nose cone fairing and vertical fin tip also appear to be Insignia White. The BuNo, "Johnsville", and Navy are all F.S. 17038 Gloss Black/Jet.

Unquestionably the most bizarre of any A-3 camouflage pattern was that seen on several VAP-61 aircraft, including BuNo 144831 during a temporary stopover at NAS Alameda on February 10, 1968. The colors consisted of F.S. 36251 Medium Gray; F.S. 36440 Light Gull Gray; F.S. 36118 Sea Gray; and what appears to be F.S. 17875 Insignia White. The pattern effectively was a wraparound scheme and included all aircraft surfaces with the exception of the aft portions of the engine nacelles.

Rarely seen VAH-4 KA-3B, BuNo 142662, wearing an olive drab camouflage that appears to match the Army's F.S. 34087. The aircraft is painted olive drab over-all, with the exception of the undersurfaces which are F.S. 27875, Semi-Gloss Insignia White. Somewhat strangely, the upper rear halves of the engine nacelles have been left unpainted, and the lower rear halves appear to have been painted olive drab! Note, too, the white vertical fin and wing leading edges.

VAP-61 RA-3B, BuNo 144847, while transient at NAS Alameda, California during May, 1961, and while wearing the rarely-seen Vietnam-era F.S. 17038 Gloss Black/Jet color scheme. This scheme was optimized for night operations and was of the wraparound variety except for the aft, hot section of the engine nacelles (which were left unpainted). Noteworthy is the aircraft fin cap, which appears to be painted F.S. 37038, matte Black. The only visible warning markings are the yellow rescue arrows.

With its drag chute fully deployed, VAQ-33 ERA-3B, BuNo 144827, decelerates following a landing at NAS Miramar, California during July, 1980. Markings are standard for type, with the upper surfaces being painted F.S. 16440 Light Gull Gray, and the lower surfaces being painted F.S. 17875 Gloss Insignia White. The unit patch is visible on the vertical fin, above the large BuNo. Noteworthy are the unfeathered propellers of the fuselage RATs and the pylon mounted AN/ALQ-76 ECM pod.

KA-3B, BuNo 138946, of VAH-10, CVW-21, then stationed aboard the USS "Hancock", while transient at NAS Barber's Point during September, 1969. Markings are standard for type, with the upper surfaces being painted F.S. 16440 Light Gull Gray, and the lower surfaces being painted F.S. 17875 Gloss Insignia White. Also as standard, the aft engine nacelle fairings have been left unpainted. Note also the Insignia White rudder and vertical fin leading edge. Sword is quite distinctive on fuselage side.

EKA-3B, BuNo 147660, of VAQ-130, CVW-19, during April, 1969. Markings are standard for type, with the upper surfaces being painted F.S. 16440 Light Gull Gray, and the lower surfaces being painted F.S. 17875 Gloss Insignia White. The unit patch, which says "VAW-13" even though the aircraft was marked VAQ-130, is visible on the forward electronic warfare systems fairing. At the time, the aircraft was assigned to the USS "Hancock".

EKA-3B, BuNo 142659, of VAW-13 and seen at NAS Alameda, California on October 23, 1968. Markings are standard for type, with the upper surfaces being painted F.S. 16440 Light Gull Gray, and the lower surfaces being painted F.S. 17875, Gloss Insignia White. The unit patch is visible on the forward electronic warfare systems fairing. Under the forward windscreen is the title, "Killer Whale" and the names of its crew. The aircraft was assigned to the USS "Bon Homme Richard".

KA-3B, BuNo 147660, of VAK-308, during immediate pre-flight activities at NAS Alameda, California. Whale fin tip markings are classic for type, with all other markings, including the F.S. 16440 Light Gull Gray upper surfaces, and F.S. 17875 Gloss Insignia White lower surfaces, being standard. F.S. 30266 (approx.) Middlestone brown nose cap is noteworthy, though periodically seen on A-3 variants. Also noteworthy is bomb bay door (starboard half, open) interior detail.

TA-3B, BuNo 144859, of VAH-123, static at NAS Barber's Point. Wraparound F.S. 16440, Light Gull Gray has high gloss finish. Noteworthy is the F.S. 17875 Insignia White vertical fin leading edge and fin cap. Also noteworthy is F.S. 17038 Gloss Black/Jet nose cap and distinctive unit patch on fuselage side. In early 1987, twenty-seven years after it had been accepted by the Navy, this aircraft was still operated by VAQ-33 to train A-3 aircrews and was regularly used for CARQUALS.

ERA-3B, BuNo 144846, of VAQ-34, while completing a training mission off the coast of San Diego, California on January 13, 1986. All markings, including the F.S. 16440 Light Gull Gray upper surfaces and F.S. 17875 Gloss Insignia White lower surfaces, are standard for type. Distinctive white rudder and elevators are noteworthy, as are the white empennage fairing (accommodating chaff dispensing equipment) and the white ventral antenna canoe.

Operated from NAS Patuxent River, A-3B, BuNo 142246, bears unusually bright F.S. 28913 Florescent Red-Orange markings signifying its research and test role. F.S. 28913 covers large sections of the aircraft's nose, vertical tail, and outer panel wingtips. Noteworthy is the F.S. 17875 Gloss Insignia White vertical fin leading edge. This aircraft now has been removed from the active Navy inventory and is owned by the Bradley Air Museum in Windsor Locks, Connecticut.

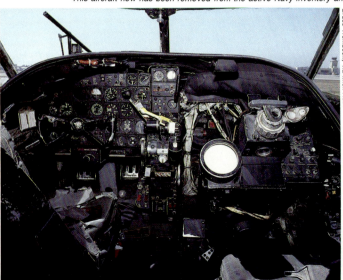 

KA-3B (left) and ERA-3B cockpit main instrument panels and starboard sub-panels. All A-3 operational configurations are fairly similar to these in terms of over-all layout and instrumentation. As can be seen, the A-3 is a single-pilot aircraft equipped with only one control yoke and one set of instruments. The sub-panels to starboard are primarily radar systems optimized to accommodate navigation needs.

Chapter 8
Tales of the Whales

Photographed in April, 1965, during its third bombing mission, NL 605 is seen dropping a Mk.82 over a North Vietnamese coastal target. It was then a tanker-configured A-3B but later became a KA-3B, then an EKA-3B, and again a KA-3B. In this last configuration, this combat veteran was still operated by VAK-208 in early 1987.

Lt(jg) Don J. Willis, a bombardier/navigator with VAH-2, kept a detailed diary from which was abstracted the story of the A-3's first bombing mission.

Providing a brief taste of life with *Skywarrior* units, this chapter is a compendium of personal recollections, sea stories, and other yarns by and about "whalers."

THE FIRST BOMBING MISSION

(The following is an excerpt from the diary of Lt(jg) Don J. Willis, a bombardier/navigator with VAH-2, CVW-15, aboard the USS *Coral Sea*.)

Monday, 29 March 1965
Vicinity Point Yankee, South China Sea
"Things have picked up since leaving Cubi on the 15th, not only for me, since I'm now in a crew flying with our new XO, Cdr. John Sundberg, and Jim Thompson, ABE1, but also for the entire Air Group. Strikes have been frequent of late—estimate half a dozen during past two weeks. Crew 2 has six 'green entry' flights logged this month as of today: the first was as part of a three-plane A-3 attack on Phu Van Army Supply Depot which was aborted due to no significant targets (this being a 'political' war, we were told not to drop haphazardly). Then there was a fighter/photo tanker job along the South Vietnam coast; two tanking missions in the Da Nang/Tom Cat/Laos Entry Point area, a weather recce over Laos and, unintentionally, (trying to raise the ship by radio) a part of North Vietnam, and the big one today—the first mission of the Heavies.

The target, Ile Bach Long Vi or Ile Nightingale, located in the middle of the northern part of the Gulf of Tonkin—60 miles or so west of Hainan Island. Three actual drop areas, designated target 1—radar site, target 4—HQ (ours) and target 5—Admin and barracks, were assigned to three 2-plane groups, led, in order, by CO, XO, & Ops. The planning was all done last night rather hurriedly since we got the final go only yesterday and half the crews flew most of the day. A rather general opinion by our clan was why in tarnation wasn't this target assigned in the first place instead of non-significant (radar speaking) ones like Phu Van—since it was the first 'test' for the Heavies and supported by Admiral Outlaw himself. I still remember what he said in his pep talk that day (19 March)—'O.K., you *Whale* people have your chance, now get out there and do a good job, we don't want any bombs out in the boonies. We could handle it when I was in the business and you can do it now.'

Back to today. It was our third time to man aircraft when the final word was 'launch'—and that they did—almost cleared the flight deck but took time in launching all *Whales* from waist cat. With four Mk 83s (1,000 lb. ea.) we grossed heavy—about 72.5 or so. Our call was *Folder Three* and the bird was NL 608 (BuNo 147660)—a tanker configured 147 series. After rendezvous and outbound to first turn point, Cdr. de Lorenzi shifted lead to us—LCdr Bob Vaughan, Ltjg Dick Toft, and Dave Lippitt, AT1, were on our wing flying NL 602 (BuNo 142649), with a load of eight 500 pounders. Radar & computer was sweet—real good gear—obtained fixes off Cap Mui Ron and an island off North Viet coast. Inbound to target experienced strong radar interference, eased up by changing mag. freq., something usually not done. The island, a mile and ¾ long by ½ mile wide finally broke out about 35 miles to go and by 30 was well in the ball park. No prominent return, unable to solve for winds—PDI erratic in close, but Cdr Sundberg did some fancy jockeying—bomb check list was completed—signaled Thompson for bomb doors as range passed 40,000 feet and was preparing for manual backup when she released close to pre-comp R/R (BIP not on)—intervalometer was set for 4 at 150—fired perfect, all 4. Our drop was at 1454. Cdr. Sundberg said—'Well, you've just become the first heavy attack bombardier to make a drop on the enemy!' A quick glance at 602 and I saw a thumbs-up in their cockpit. They headed for Cubi. Shortly thereafter, the CO and Morgy were coming over target.

After LCdr. Richards & Terry tanked CAG, we topped them off and headed for *Mustang*. Meanwhile the A-4s were having their hands full. Heavy ground fire punctured two gas tanks on the first pass; Cdr Mongilardi, CO of VA-153, was hit bad and streaming fuel. Through help of *Royal Lancers*, they 'rendezvoused' with our tanker—the CO was at 600 lbs—Richards tanked him all the way to the ship and the deck was cleared for a straightin. We were recovered by this time and taking a hot refuel when Cdr Mongilardi trapped. He had complete utility hydraulic failure in addition to the fuel leak, numerous holes from flak and machine gun fire was in the aft fuselage area near the tailpipe; his aircraft also appeared scorched, as by fire.

We were catted again and took a station overhead while CO and crew went out 100 miles. We tanked a couple of *Corktip* photo F-8s and bored holes rest of the time. *Mustang* gave 601 (CO) a *Charlie* and had them pass extra fuel to us—to make our total 20 thou; about 5 minutes later 'your signal *Charlie*' was given us; we all laughed as we could have stayed up there for hours.

"The XO made two good traps today—little high on first. Collected bomb-arming wires and took them to the guys in the ASB shack and thanked them for coming thru with good gear. XO and CO attended debriefing. At chow discovered our losses were more than we'd heard. Cdr. Harris, CO of VA-155 was shot down, but picked up by submarine, *Charr*, south of target. However, as of yet, no word on the CO of VF-154 (F-8s), Cdr Donnelly or of LCdr Hume, also of 154. They also lost another bird day before yesterday on a strike on the same island... Lt. Buck Wangeman, pilot, was rescued. Double-nuts, 400, with CAG's name on it, was the unfortunate bird.

Matt Gardner, squadron AI officer, was just in from air intelligence; said the SPAD RESCAP reported that three hours after the raid, called 'Rolling Thunder 8-FOXTROT,' the island was still exploding. Apparently there were a few tunnels or underground ammo stowage. Our radar site target is reported destroyed. Matt says a couple of strings of bombs were within 200 ft of HQ—not bad for ballparking, but no bullseye.

Its 0130, another SPAD was launched. RESCAP will be working long searching for those lost pilots—there goes another one—especially since a beeper has been heard about 20 miles from the island.

Tonight, Cdr. de Lorenzi answered the phone in the ready room with 'Bomber-Squadron Two.'"

THE INSIDIOUS ENEMY

(Reprinted from an article in the January, 1965 issue of *Approach*, the monthly magazine published by Commander, Naval Safety Center, NAS Norfolk.)

"On a day IFR flight on 28 June 1964, an

85

RA-3B (BuNo 144829) of VAP-62 departed NAS Jacksonville for the Northeastern United States. An intermediate passenger stop was scheduled to drop off the photo-navigator to attend a survival school. Pilot, photo-navigator and photo technician comprised the crew. At the passenger stop, the photo-navigator departed the aircraft and the photo technician moved into his position. This was temporary since another photo-navigator was to be picked up later at the destination to return with them to home base.

After dropping the original photo-navigator passenger, the aircraft took off at 1932Z. At 1935, the pilot made initial contact with the Air Route Traffic Control Center. The Center cleared him to destination on a preassigned frequency of 279.6 mcs. Clearance was acknowledged and the pilot gave his altitude as passing 22,000 feet. The pilot asked the Center to repeat a transmission at 1937:30, then to 'wait one'. Between 1944 and 1947, five calls by the Center produced no reply.

At 1949, the pilot replied negative when the Center asked if he was in any difficulty. To a query on altitude he reported he was level at flight level 310. At this time, he was assigned a new frequency, 263.1. He acknowledged 263.1. The Center tried contact several times, requesting him to 'squawk ident' on his transponder on the new frequency but there was no response.

At 1953, the Center's radar showed the aircraft making figure eight patterns. With communications still lost with the RA-3B, a Flight Service Station was instructed to attempt contact.

At 1955 the pilot was contacted by a second Flight Service Station on 272.7 mcs when he was heard attempting to contact the Center on this frequency. He informed the Station that he was on J-32 at flight level 310 and was not sure of his position. The Station lost radio contact with him at this time. The Center regained radio contact with him a minute later on his assigned frequency.

Communications were regained between the Center and the aircraft for the next 18 minutes. During this time, positive radar identification was effected and a radar steer given the pilot. He reported navigational lock-on at 2005 and was resuming his own navigation. The Center acknowledged and the pilot thanked the Center for their patience.

At 2014 the Center effected a radar handoff and notified the pilot, directing him to contact his new Center on 385.6 mcs. The pilot was extremely confused and was unable to read back the frequency. This was his final transmission. The Center repeated its instructions. There was no reply...

At 2021 the aircraft made a wide sweeping left turn. At 2033 a TF-9J was diverted to attempt an intercept. The TF-9J pilot found the RA-3B at flight level 410, still climbing. Upon reaching flight level 440, the RA-3B engines stopped emitting contrails indicating that both engines had flamed out. The aircraft then commenced a rapid left roll and went into a vertical dive at a tremendous rate of speed. At 15,000 feet the aircraft righted itself to relatively level flight with a steady descent of 2000 feet per minute. At 13,000 feet the Cougar pilot joined up on right wing and observed that the person in the right seat—the photo technician—was slumped and apparently unconscious. The RA-3B continued to descend and crashed in a small wooded area and on into a cornfield...

Ironically, accident investigators reported that 'the crew's section of the cockpit area remained in remarkably excellent condition and was considered to be survivable.' The pilot and crewman received minor cuts and abrasions, none of which could have caused fatal injury.

Medical examination indicated a high probability that the pilot and crewman had died prior to impact as a result of a prolonged state of hypoxia.

The pilot was found wearing his helmet with his oxygen mask dangling from the left retention fitting. The right seat crewmember's oxygen mask was on the floor of the cockpit in front of him. His helmet was behind the pilot's seat. Both men's oxygen hoses were routed through the channel of their torso harness and connected to the aircraft oxygen system. The converter still had lox remaining. The oxygen system switches for both the left and right front positions were in the off position."

The most probable causes of the accident were partial loss of cabin pressurization and failure of the crew to use oxygen as required by NATOPS.

THE SAGA OF GIGI

(Reprinted from an article by Jug Bernhardt in the July 19, 1974 issue of *Crosswind*, the base newspaper of NAS Whidbey Island.)

"A cute and shapely lass, a whole carrier full of lonely and lovesick sailors, and a vast ocean. With these ingredients on a nippy winter evening long ago in Alameda, California, a chain of events began that would endure for 17 years.

The shapely lass was the *Golden Goddess*, a statuette standing 43 inches tall and measuring a petite, but eye-catching 26, 19, 24. She wore nothing but a gleam in her eye and a coat of golden paint. The carrier full of lonely and lovesick sailors was the USS *Hancock* and the men of Det Delta of VAH-4. And the ocean was the Pacific.

This set the stage for Clifford Hornung, then attached to Det Delta, to attempt the perfect statue-snatching of the centure. The *Golden Goddess* resided in *Tim's*, a nightclub in Alameda. 'Why should all those men on that big carrier be without a mascot?' Hornung thought.

Then the fun began. Chief Hornung made off with the statue and as the year turned into 1958, the *Golden Goddess* deployed aboard *Hancock* to the Western Pacific. The *Hancock* and VAH-4 Det Delta arrived on station and immediately a cry of envy rang from every ready room in WestPac because Det Delta had *Gigi* (her new, abbreviated name).

Detachment Charlie of VAH-4 was stationed aboard the USS *Shangri-La* and they readied themselves for a night penetration into the abode of *Gigi*. When the two carriers were in a Japanese port at the same time, the dastardly abduction was carried out.

Det Charlie had her, but not for long. A midnight boarding party was organized and the pirates of Det Delta stormed the *Shangri-La*.

When the attack was over, the *Golden Goddess* was back home with her original admirers. But as the laurels of victory are always gained, Det Delta lost one of its less prized members; an Ensign was left behind.

Det Charlie offered the Ensign in exchange for the cherished *Gigi*, but the barter was promptly rejected. Sea stories have it that the Ensign was later returned to Det Delta with a rose between his teeth and wearing nothing but a coat of fresh gold paint.

In the next sequence of events in the life of the *Golden Goddess*, she won her wings by flying with the Navy.

While Det Delta was temporarily shore based at NAS Atsugi, another squadron, VAH-2, got wind of the coveted lass. By removing hinges and doors to the Det Delta maintenance spaces were *Gigi* was stored, the heavies from VAH-2 stole the golden girl from her owners. She then boarded an A-3B and made her way to NAS Cubi Point. Heavy Two was attached to the USS *Midway*, and in February of 1959, they proceeded to NAS Naha, where they were debarked.

VAH-8, also attached to the USS *Midway*, learned that *Gigi* had been taken to Okinawa. An attack was readied: they planned to sent an A-3 and crew to NAS Naha and abscond with *Gigi*.

The plans were carried out and, as the Heavy Eight *Whale* arrived at Naha near sunset, intelligence reports were received that *Gigi* was in the bomb bay of one of the aircraft assigned to VAH-2.

The inky darkness of early morning still remained when the VAH-8 crew removed the sexy statue from her vulnerable hiding place and escorted her to their own aircraft. Before sunrise they were winging their way to the USS *Midway*.

The *Midway* was then steaming back to the 'states' and NAS Whidbey Island where the *Golden Goddess* was to occupy a position of honor in the duty office of Heavy Eight.

October of 1959 brought three burly enlisted men from VAH-6 to the VAH-8 duty office and the subsequent forcible snatching of the goddess. Heavy Six then deployed in February of 1964 with the goddess aboard the USS *Ranger* on that ship's first WestPac cruise.

Her fame and glory reached every ready room and bar from Singapore to Sasebo. Her history became a highly coveted item and all squadrons who had possession of her had a tough time retaining her.

"The following is an account of the squadrons, ships and sailors that have had custody of *Gigi* as recorded in the *Golden Goddess* scrapbook kept by VAQ-135.

During an inport period at NAS Cubi Point, the men of VAH-6 decided that the sex starved sailors of WestPac needed a little entertainment. One evening they took *Gigi* to the Cubi Point Officers Club. She was promptly stolen by a group of officers from a destroyer that was in port at Subic Bay. Reacting instinctively, the Heavy Six contingent went to the destroyer and stole the ship's bell.

A trade was effected and each party returned with its original possession. The implication of the destroyer incident was obvious to the men of Heavy Six: extra precautions would have to be taken if they were to retain the *Golden Goddess*. The goddess was subsequently bolted to the floor of the Heavy Six ready room aboard the *Midway* and the heads of the bolts were removed. It was thought that this would surely thwart all future theft attempts. This was an incorrect assumption.

Late one evening a fighter squadron stole into the ready room, tied the Assistant Duty Officer and used a blow torch to remove *Gigi*. VAH-6 remodeled the goddess after recovery (her legs had been slightly damaged by the fighter squadron) only to have her stolen by VAH-8 who brought her back to the 'states' in early 1965.

In the confusion of VAH-8's homecoming, several members of VAH-4 saw the goddess placed in a locker in the VAH-8 duty office. A VAH-4 officer posed as a VAH-8 duty officer, obtained the key to the locker and had a duplicate made. He returned the keys to the duty officer within the hour assuming she had gone undetected. That night a raid was attempted. Access was obtained by means of a roof top entrance. When the locker was opened, all that was there was a note which said 'Sorry about that'.

In April of 1965 an aircrewman from VAH-4 Det Golf, who at the time, had a broken arm, 'shanghaied' the statue by using a pair of lock cutters. In May of the same year, Det Golf deployed aboard the USS *Oriskany*. *Gigi* survived the tragic fire aboard that ship and returned to Whidbey in December of 1965. In November of 1966, Det Golf again deployed to Southeast Asia, with enroute stops at Hawaii, Wake, Guam, and the Philippines. The detachment left Whidbey with *Gigi*. Things went smoothly and *Gigi* completed her second combat cruise aboard the USS *Oriskany* and returned to Whidbey Island in January, 1968.

In the ensuing months, several kidnap attempts were made by various A-3 squadrons, but all were in vain. It was not until March of 1971 that VAQ-130 Det 3, then deployed on the USS *Oriskany*, was able to take possession of the prize beauty.

In October of 1971, the *Oriskany* was in Cubi Point preparing to return to the 'states'. In the excitement of returning home, *Gigi* was ignored for a few moments, just enough for an aircrew of VAQ-208 to snatch her. The VAQ-208 reserves were based at Cubi Point and were flying mail and aircraft parts to the USS *Enterprise*, which was stationed in the Indian Ocean. VAQ-208 returned to NAS Alameda in February, 1972, and the *Golden Goddess* was placed in a glass enshrouded pedestal in the Commanding Officer's stairwell in the VAQ-208 hangar.

On April 6, 1972, a small band of men from VAQ-135, posing as members of a Public Works team from Disease Vector Control, were able to clear the VAQ-208 hangar under the guise of a mandatory fumigation. They leisurely removed *Gigi* from her glass prison and return her to active duty.

VAQ-135 now has possession of the *Golden Goddess*; holding the shady lady who has been the object of intrigue and mystery as she drives men to do pretty unusual things just to have her company."

THE SHREWD TAMING OF A CAT
THE ROTA'S BLUE BEAR
AND
A BEAR'S SURPRISE PARTY

(Contributed by Lts. Mike Dengler, Ed Kane, Mike Laviano, and Randy Masters, the following sea stories were told not in a bar but in the forward wardroom of the USS *Kitty Hawk* where, perforce, liquid consumption was limited to coffee and soft drinks. Hence some credibility may be attached to these stories...)

If one believes that life in a FAIRECON is all SIGINT, ELINT and other spooky stuff, one may be surprised to learn that there is more to life than mere electronic wizardry. Animal husbandry and airborne partying appear to have sneaked in insidiously in the evolutions if the following yarns are to be given credence.

The Shrewd Taming of a Cat: Aware that a crew was to ferry a *Whale* from NAS Alameda to NS Rota, the wife of a senior officer prevailed upon the kindness of the crew in arranging for her prize kitten to be flown to Spain. As sedation plays havoc with the precious coat of purebred cats, the crew was duly instructed not to have the kitten drugged. Put untranquilized in the confines of an animal travel crate, the kitten voiced his displeasure immediately upon departure from the West Coast. By the time the aircraft left the East Coast after a short refueling stop, the hapless feline was howling and the crew's patience had been stretched beyond reasonable limits. Motivated by either self-preservation, urging of other members of the crew to 'do something,' or pity for the tormented kitten, an aircrewman—claiming to know how to appease animals—bravely volunteered to take the cat out of his crate and to calm him down on his lap.

Dutifully proceeding with his charitable intention, the *Batman* literally let the cat out of the bag. One of Murphy's Laws immediately went into force. In true jack-in-the-box fashion, the kitten sprang out of his crate and, clawing at everyone in his way, clambered all over the cockpit. Fearful for the 'safety' of his crew, and once again proving that naval aviators are fast thinking, the skipper swiftly devised an emergency procedure for a situation not covered in NATOPS. Instructing his crew to go on oxygen, he depressurized the aircraft. Hypoxia soon overcame the kitten who was then recovered. His subsequent fate remains unconfirmed. Human ingenuity had prevailed over animal fury: the *Skywarrior* and its crew made a safe landing at Rota.

The Rota's Blue Bear: Keeping harmonious relations with the host nation is a collateral duty when based overseas. Thus, when it was learned that the city of Rota wished to add a Russian blue bear to its small zoo's collection, VQ-2 was prompt to do its share for the benefit of diplomacy. A crew on detachment to Incirlik was to locate one of the rare bears in Turkey and bring it back to Rota. After all, the *Batmen* had already proven to be expert 'animal handlers.' With the help of local hunters, an expedition set out in the wild and, taking to heart their diplomatic mission, the *Batmen* succeeded in trapping a bear (catching a member of the Ursidae family, not landing a Tupolev Tu-142 aboard a carrier!). Placed in a crate for the trip back to Incirlik, the bear was then readied for the flight to Spain.

Experience coming in handy, the crew obtained an ample supply of tranquilizers and doped the bear for his air voyage. So far, so good! The *Whale's* crew and its precious beast made it safely to Athens. During the stopover in Greece, when the slowly awakening bear was chained to the *Skywarrior's* nose gear to answer a call of nature, tragedy almost struck. Believing that the shaggy animal was a dog, and blaming the naval aviators' lack of sensitivity for the plight of their unattended ward, an Air Force nurse went to play with him. Now fully awake, the bear took an unnatural liking to the airwoman and nearly mauled her! Cursing the lack of zoological knowledge of the female "blue suiter," the *Batmen* had a difficult task controlling and sedating their unruly ward. Finally, with flight plan filed and bear tranquilized, the aircraft departed for Rota. Met by officialdom upon arrival, the crew handed over the bear to the *alcalde* (mayor) of Rota. Photographers were on hand to record the nice deed and one of their pictures appeared in the *Navy Times*. Cropped from the waist down, the photograph shows contented naval aviators and proud *alcalde* smiling, and the tearful face of the *alcalde's* young daughter. The full-length photograph provides the clue to the senorita's tears. The no-longer sleepy bear was chewing on her hand! The moral of the story: do not expect Russian blue bears from Turkey to be reared on the fine points of diplomacy.

A Bear's Surprise Party: Although the *Whale* was adapted to fulfill numerous roles not originally planned, our valiant gal had a difficult time when attempts were made to use her in the intercept role. This unusual task came about when Tupolev Tu-142 *Bears* operating from Cam Ranh Bay regularly probed U.S. defenses in the Philippines. At the time, no fighters were based at Cubi Point, and VQ-1 and its EA-3Bs were given the task of showing the Soviet patrol bombers that their antics were not going unnoticed. On several occasions EA-3Bs were sent on "BearCAP," but a combination of events prevented successful interception. For the crews, without the incentive of *Bear*-watching, these sorties were long, tedious and unrewarding. Morale was low and *Bear* patrols were dreaded exercises. Wishing to add spice to an otherwise bland assignment, a young skipper handed sealed, classified-looking envelopes to his crew, and sternly warned them not to open the envelopes unless properly instructed to do so. The EA-3B was launched and when monotony was dulling senses the skipper ordered the crew to open the envelopes. To their surprise, the *World Watchers* discovered non-regulation party favors and other paraphernalia. Soon, the party was on and a good time was had by all.

Then it happened! A *Bear* was spotted. Wasting no time, the crew got in position and did its work. Everything was going by the book when someone noticed that there were more puzzled Soviet faces looking at the American warplane than was usually the case. The alert crew soon solved the mystery: the *World Watchers* still had their birthday hats perched atop their flight helmets! Until the publication of this story, it is doubtful that the Soviet flyers have been able to come up with a logical explanation for this new capitalist treachery...The mid-

In the midst of their deadly serious electronic intelligence gathering mission, the "Batmen" of VQ-2 managed to keep a well-developed sense of humor as evidenced by the stories told here. Devoid of unit markings, this EA-3B from VQ-2 was photographed over the Mediterranean Sea on May 15, 1970.

Halfway around the world, the "World Watchers" of VQ-1 (TA-3B shown) have an equally developed sense of humor. One of their pranks, as recounted here, must have caused some consternation and more than a little wonderment to Soviet Naval aviators.

87

night oil must have burned at the headquarters of the *Aviatsiya Voenno-Morskovo Flota* (AV-MF, Aviation of the War Fleet)!

THE FLYING SPORTS CAR

(Contributed anonymously by one who knows "what evil lurks in the hearts of men.")

An individual in the A-3 community had to move his car, an MG-TD, from Sanford, Florida, to California. Somehow he talked two others into helping him transport it in the bomb bay of an A-3B. With a fair degree of ingenuity and some hard work, they surreptitiously hoisted the MG-TD into the bay and even got the doors shut. The flight was uneventful until arrival over NAS Alameda. However, when it came time to lower the gear, nothing happened. The handle in the cockpit moved, but the gear stayed in place.

Deciding that something was slightly abnormal, the crew left Alameda and proceeded out past the Golden Gate to diagnose the malfunction. Finding that normal hydraulic means were unavailable, the B/N dropped the inner hatch leading to the companionway to get to the emergency landing gear controls. Well, Murphy's Law prevailed; in spite of its "never fail" reputation, the blow-down pneumatic secondary system did not yield the desired result. One option remained. However, it required that a crewman get to the gear access hole in the rear of the bomb bay, the very bay occupied by the MG-TD. A short conference in the cockpit produced the unanimous decision that if the bird was going to belly land at Alameda, it was not going to have a certain English sports car in its belly for the accident investigators to raise their eyebrows over.

The B/N went back to the bomb bay and, shades of Slim Pickens in *Dr. Strangelove*, proceeded to cut away the lines restraining the MG-TD until it rested on the bomb bay doors. Just as he re-entered the companionway en route to the cockpit, the aircraft shook with a resounding "ka-chunk" as the main gear finally came down! Without further ado, the *Whale* crew landed at Alameda with the MG-TD intact and the world none the wiser about the sports car that was almost jettisoned over the Farallons...

THE VAH-5 MUSHMOUTH INSIGNIA

(Story abstracted from a letter sent on November 10, 1968 to the Chief of Naval Operations by the Commanding Officer, Reconnaissance Heavy Attack Squadron Five.)

Shortly after moving to NAS Sanford, Florida, in 1955, Composite Squadron Five (the forebear of VAH-5/RVAH-5) adopted the nickname "Savage Sons of Sanford" and a new insignia to reflect the name of an earlier Operations Officer, Cdr. Savage, and the fact that it was equipped with the North American AJ-1 *Savage*. The insignia consisted in a stylized black face with a wide mouth and a bone through a topknot. Although originally coined by an outsider as a derogatory term to describe the new insignia and, by implication, unit personnel of VC-5, the nickname "Mushmouth" caught on and was proudly used by the squadron.

All was fine for several years. Then, as the CO described events:

"In April 1962, VAH-5, embarked in Forrestal, was participating in a Presidential sponsored cruise in honor of Ambassadors from the African States. The then Chief of Protocol, Angier Duke, was aboard. Fearing the Mushmouth insignia might be offensive to the guests, he directed the removal of all "MUSHMOUTHS" from VAH-5 airplanes, spaces and people. To the best of our knowledge, never before in the annals of naval history has a command been so suddenly stripped of its insignia. It was a move which created much ill feeling among present and past "MUSHMOUTHS"... In September 1962 the squadron, although fearful of offending the American Indian, adopted the new insignia... and retained the nickname "Savage Sons of Sanford". The present insignia, the "one-eyed Indian" (author's note: a profile of an Indian head with full feather headgear, over an arrow), is a good insignia, and is symbolic of the Savage Son theme; but it is not an insignia with which the troops can identify—like the "MUSHMOUTH"!

Later in his letter to the CNO, the CO requested authorization for the adoptions of a revised "Mushmouth" insignia. The authorization was not granted, and RVAH-5 used the Indian head insignia until it was decommissioned on September 30, 1967.

Among the "Whale's" many contributions to U.S. Naval Aviation, none is more important than the sterling work performed by tanker crews from Heavy Attack Squadrons and Tactical Electronic Warfare Squadrons during the war in Southeast Asia. This life and aircraft saving task is exemplified by this view of a tanker-configured A-3B from VAH-4 Det Mike refueling an F-4B from VF-96 over the Gulf of Tonkin during the 1965-66 deployment of CVW-9 aboard the USS "Enterprise" (CVAN-65).

Chapter 9
A-3 Technical Description

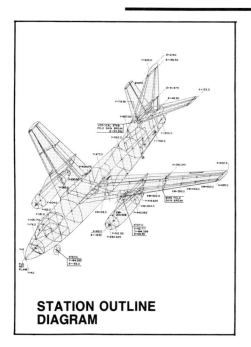

STATION OUTLINE DIAGRAM

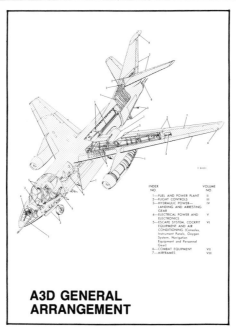

A3D GENERAL ARRANGEMENT

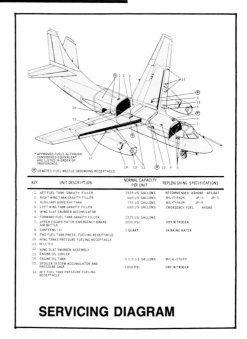

SERVICING DIAGRAM

The following description applies principally to the KA-3B version of the *Skywarrior* subsequent to the removal of its bombing equipment and tail turret. The powerplant installation and the airframe are similar for all A-3 versions, except for crew accommodation and aft compartment arrangement—which in the case of the EA-3B, ERA-3B, NRA-3B, RA-3B, TA-3B, and VA-3B include a pressurized compartment in lieu of the bomb bay.

The KA-3B is a three-place, twin-engined, swept wing monoplane equipped for air refueling operations. Designed for carrier operations, it utilizes catapult assistance for takeoffs and conventional arresting gear for deck landings. The aircraft is equipped with a retractable tricycle landing gear, a drag chute, and a steerable nosewheel.

Skywarriors modified by NARF Alameda into KA-3Bs have included aircraft from different production lots. Hence, three types of KA-3Bs are generally similar with differences confined to fuel tank arrangement, fuel transfer systems, and wing leading edge configuration (with and without CLE). Type I and Type II KA-3Bs are late production airframes with CLE wings, Liquidometer fuel quantity indicating system, and a six-step fuel balance system using forward to aft fuel transfer pumps. Type III KA-3Bs are older airframes with standard wings and Avien fuel balance and quantity indicating system using gravity transfer from forward to aft fuel tanks. Difference between Type I and Type II aircraft are limited to instrument and console panels.

Principal dimensions of the KA-3B, as quoted in the July 15, 1970 issue of the NATOPS Flight Manual (NAVAIR 01-40ATE-1) are as follows:

Length (fin erect and probe installed)	78 ft.	(23.77 m)
Length (fin folded— probe not installed)	74 ft. 6 in.	(22.71 m)
Wing span (wing spread)	72 ft. 6 in.	(22.10 m)
Wing span (wing folded)	49 ft. 5 in.	(15.06 m)
Height (fin erect)	23 ft. 6 in.	(7.16 m)
Height (fin folded)	15 ft. 9-3/8 in.	(4.81 m)
Maximum height during wing folding	27 ft. 3-3/4 in.	(8.32 m)
Height over wings (wings folded)	16 ft. 8 in.	(5.08 m)
Wing area (Cambered Leading Edge)	812 sq. ft.	(75.44 sq.m)

Fuselage—The fuselage structural arrangement is simple, straightforward and generally conventional. Built in three sections, nose, center, and tail cone, it is constructed mainly of aluminum alloy materials embodying a high percentage of 75ST material. A combination of longitudinal and traverse skin stiffeners is used, with the stiffeners located as dictated by the need to provide miscellaneous cutouts. Two heavy keels run from the bomb bay almost to the tail of the aircraft, forming the sides of the bomb bay and picking up catapult fittings as well as the arresting gear fitting strength members at the aft end.

Ahead of the cockpit, most of the nose is taken up by navigation radar antenna and related electronic equipment. The air conditioning equipment is installed on the right side, below the cockpit floor. Immediately behind this compartment and under the forward fuselage tank, there is an electronics compartment on the right side and the aircraft's basic power system on the left. This power system consists of hydraulic pumps and two DC and two AC electrical generators driven from auxiliary drive units which are, in turn, turbine driven from compressed air bleed from the jet engines. An auxiliary tank is located above the original A-3 bomb bay, with the Hose Reel Unit being mounted in the aft portion of that bay. The external fairing for the refueling drogue is mounted beneath the center fuselage, between the main landing gear. The aft fuselage tank is fitted astride the main landing gear bay while the rear fuselage section contains more avionics, the 20-liter oxygen converter system, and the drag chute compartment. To this latter section are attached the speedbrakes (both sides), the fuel vent mast (left side only), the JATO mounting hooks (both sides), and the arresting hook (beneath and on the center line).

Crew Accommodation—The crew consists of pilot, navigator, and plane captain in a pressurized cockpit.

A cockpit pressure of 5,000 ft. (1,525 m) is maintained up to a flight altitude of 13,500 ft. (4,115 m); above that point, a cockpit pressure differential of 3.3 psi over flight pressure is maintained. For combat operations above 36,500 ft. (11,125 m), the 3.3-psi pressure differential is progressively reduced until a 1.3-psi differential is reached at 48,000 ft. (14,630 m).

The pilot, in the left-hand seat, and the navigator, seated to his right, face forward, with the plane captain facing aft, back-to-back with the pilot. The seats for all three crew members are standard, non-ejectable. The pilot's seat is electrically-adjustable, whereas the navigator's set is manually-adjustable and that of the plane captain fixed.

Normal ingress and egress on the ground is through the ventral door of the emergency escape chute, with foot- and hand-holds provided on the smooth surface of the chute. When extended together, the inner and outer escape chute doors serve for emergency exit of the aircraft, on the ground or in flight. In an emergency, the bottom fuselage skin (outer portion of the escape chute) is power operated through cartridge-fired cylinders and remains open up to the aircraft's design limit speed; it provides a wind screen for the crew during bailout. The inner door, which forms the floor of the cockpit and serves as a companionway for access to the bomb bay, is pulled down through

Designation and Bureau Number as painted on the aft fuselage side of KA-3B. This aircraft is currently assigned to VAK-208.

The A-3's cockpit canopy transparencies and associated framing have gone through a variety of configuration changes throughout the aircraft's life. Illustrated upper left is the canopy for the XA3D-1; in the upper right is the canopy for the TA-3B; in the lower left is the canopy for the EKA-3B; and in the lower right is the canopy for the RA-3B. The latter aircraft is seen being modified at NARF North Island.

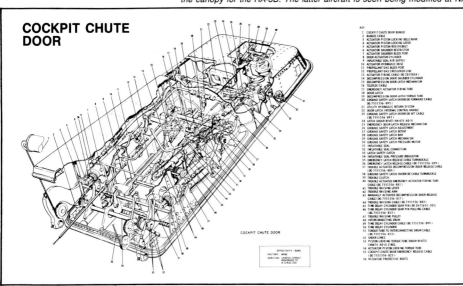

The main crew entrance hatch is located immediately to the rear of the nose gear well and just ahead of the bomb bay. It is hinged at its forward edge, only, and is provided with steps to facilitate ingress and egress. Variations to this hatch configuration can be found among the various A-3 models.

a mechanism tied to the lower power door. The cockpit must be depressurized in order to open this door. The chute is completely confined with smooth walls to effect a safe exit. An aft-sliding rectangular hatch in the canopy above and between the seats of the pilot and the navigator serves as the primary escape route in the event of ditching and as an alternate bailout route. Normal procedure calls for this upper hatch to be left open during takeoff and landing.

In the EA-3B, ERA-3B, RA-3B, TA-3B, and VA-3B versions, the inner door is deleted to provide direct access to the aft cabin; the lower (outer) door is thicker as it serves to seal the pressurized cabin and cockpit. In the EA-3B, TA-3B, and VA-3B versions, an additional inward-opening upper escape hatch and a right side outward-opening escape door are provided for use by the crew members in the aft compartment. In the ERA-3B an upper escape hatch has been added over the new ECM operators' compartment.

Wing Surfaces—Of all-metal construction, with two-spar torsion-box type structure, the cantilever monoplane wings are mounted high on the fuselage. Wing sweepback is 36 degrees at the quarter-chord, dihedral is nil, aspect ratio is 6.75, and wing thickness tapers from 10% at the root to 8.25% at the tip. The outer wing panels fold upward hydraulically to reduce span for carrier stowage from 72 ft. 6 in. (22.10 m) to 49 ft. 5 in. (15.06 m). Lug-like fittings, which are required at the folding joint to accommodate the hinge and locking pins, are made integral with the upper and lower wing spar caps. Warning flags, resembling red beer cans, project below the surface of the wing as visible pre-flight check to warn when either the wing locking pins or gust locks are engaged.

NACA slotted flaps, which pivot about external supports, are mounted on the trailing edge inboard of the folding joint. Total flap area is 82.4 sq. ft. (7.65 sq m) and the flaps can be depressed a maximum of 36°. Lateral control is provided by a combination of ailerons and spoilers. Two independent power boost systems for the ailerons are incorporated and there is a manual backup system for emergency field landings. Aileron area is 51.4 sq. ft. (4.8 sq m) and aileron travel is 21° up or down.

COCKPIT ENTRANCE AND EXIT PROVISIONS

CREW MOVEMENT AND COMPARTMENT DIAGRAM

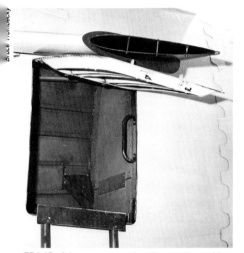

ERA-3B aft baggage compartment serves to provide space for crew member's travel bags and related transportables. It is located underneath the fuel dump.

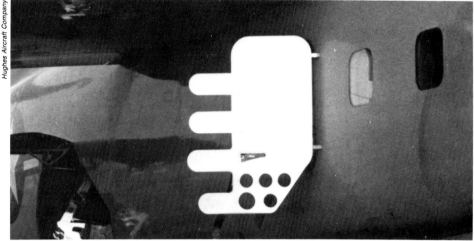

The TA-3B is equipped with a starboard fuselage side mounted door that also is designed to serve as an air dam during emergency or pre-planned airborne egress. It is equipped with aerodynamic baffles and plates which help disrupt the buffeting that occurs when the door is opened in flight.

Wing spoilers, which at high speed approximately double the rate of roll produced by the ailerons alone, are located on the upper surface of the wing just forward of the wing flaps and at the outboard end of the inner wing panels. The spoilers operate in conjunction with the ailerons whenever the aileron power control system is in operation; otherwise the spoilers are automatically faired. Spoiler travel is 58° from the faired position in the up direction only. Automatic leading edge slats[1] are provided and, for aircraft fitted with cambered leading edge, are built in four sections: one between the fuselage and the nacelle, one between the nacelle and the folding joint, and two on each outer wing panel. On aircraft not fitted with the cambered leading edge, the slat section between the fuselage and the nacelle is deleted. A fuel tank is located in the inboard section of each wing, between the spars, as an integral part of the wing structure.

Tail Surfaces—Of all-metal cantilever monoplane type, the tail surfaces are swept. The horizontal tail surfaces, which have a span of 25 ft. 5-3/8 in. (7.76 m), an area of 167.43 sq. ft. (15.55 sq m) and 10° dihedral, include an all-moving stabilizer; elevator travel is 30° up and 15° down. The rudder has 30° left and right travel and incorporates a tab with 15° travel to each side. The rudder and elevator controls are hydraulically boosted with manual backup provided for emergency use. To enable the aircraft to fit on the hangar deck of carriers, the vertical surfaces fold down to the right, reducing fin height from 23 ft. 6 in. (7.16 m) to 15 ft. 9-3/8 in. (4.81 m). A ground-

[1] An odd (and scary) habit of the A-3 involved the outboard wing slats. If, for some reason, the outboard wing slats failed to open, and it was not noticed by the crew, the aircraft went through an uncontrollable pitch-up as it slowed to normal flap and gear speed. With stuck slats, it was necessary to maintain an approach speed 25 to 35 kts above normal slats and flaps out approach speed.

selectable valve enables the vertical tail surfaces to be folded with the wings, or left up and locked. The vertical surfaces can be folded only if the wings are also folded.

Speedbrake System—Single-hinged speedbrakes, with an area of 39.46 sq. ft. (3.67 sq m), are located on both sides of the aft fuselage. They are hinged at the forward end and are operated by hydraulic cylinders. Controlled by the pilot, the speedbrakes have only two positions, fully closed or fully opened (45°).

Landing Gear—The steerable nose gear retracts forward and is fitted with a 32 x 8.8 high-pressure tire. The main gear retracts sideways and rearward into fuselage recesses beneath the rear fuselage tank; the main wheels are fitted with 44 x 13 high-pressure tires and a Hytrol non-skid braking system. Wheel track is 10 ft. 5 in. (3.13 m) and wheelbase is 26 ft. 8-7/8 in. (8.15 m). The nose gear steering control is a "pistol-grip" located on the port side of the pilot's instrument panel. A tail bumper, which comprises a sheet metal yoke carrying a small solid rubber tire, is fitted beneath the rear fuselage, just ahead of the arresting hook. Normal landing gear operations are hydraulic but an emergency pneumatic system is also fitted.

Drag Chute System—A drag chute measuring 24 ft. (7.32 m) in diameter is provided for safe landings on wet runways and during instrument landings made under GCA conditions or emergency landings made at high gross weight. The chute is for landing conditions on land only and is not intended for shipboard use. The system is designed for a normal touchdown speed of 150 knots (278 kmh) and an emergency speed of 170 knots (315 kmh). Housed in the tail section compartment of the aircraft, the chute is deployed after touchdown and is actuated by a single electrical switch

controlled by the pilot. The control system has safety features to prevent premature and accidental release of the chute and to permit jettisoning of the chute in an emergency.

Catapult Equipment—The catapult hooks are located immediately forward of the bomb bay doors. Extension is accomplished by manually releasing the mechanical uplatch, while retraction is accomplished hydraulically as the landing gear is raised. The catapult holdback fittings are located at the bottom center of the fuselage and at the aft end of the main gear well.

Arresting Hook—Of conventional design and steel construction, the arresting hook fits in a recess beneath the aft fuselage. The hook is

Cont. on p.108

COCKPIT ENCLOSURE

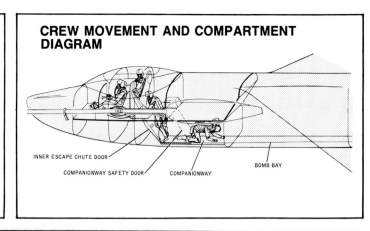

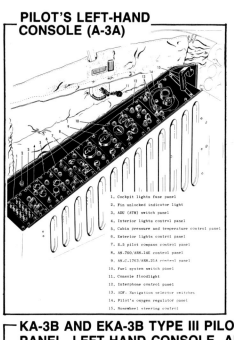
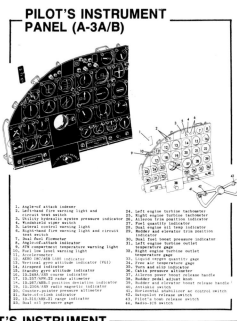
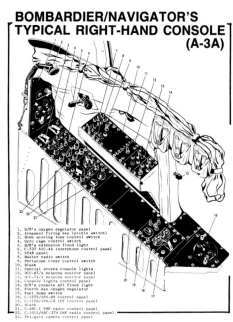
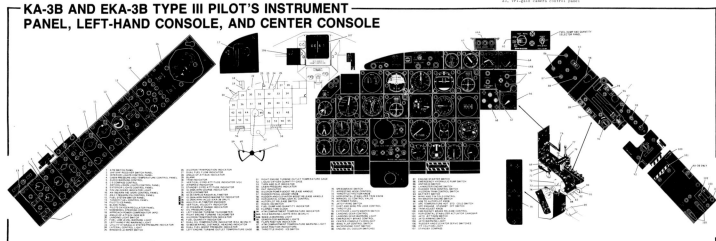
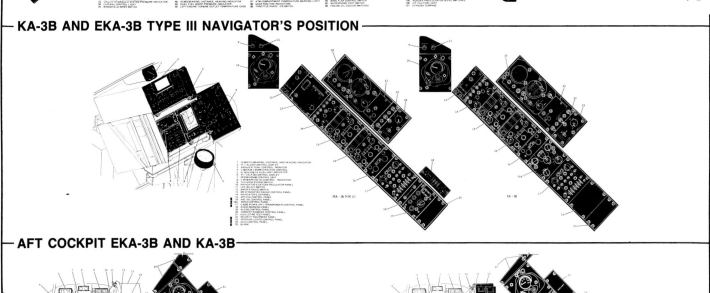
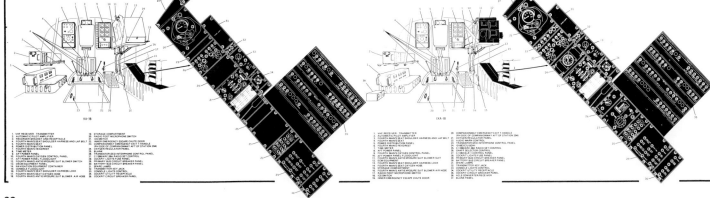

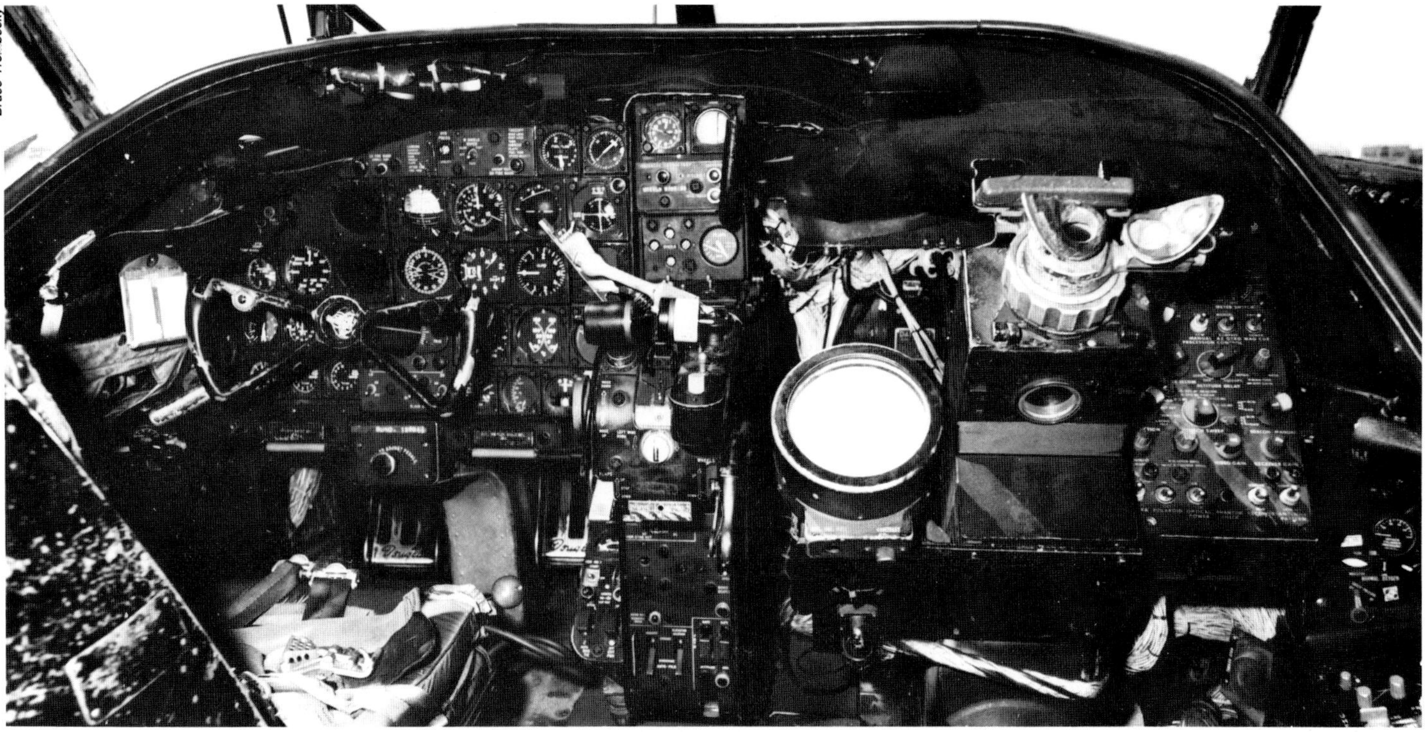

Forward instrument panel of a KA-3B. Flight instrumentation generally is similar to that found on other A-3 configurations. Primary forward cockpit panel variation is seen on the left, with navigation and communications panels occupying the space normally serving to accommodate the bombardier. Radar scope to left of station is optimized for use during rendezvous for refueling operations. Throttle quadrant can be seen on top of center console.

Bruce Trombecky

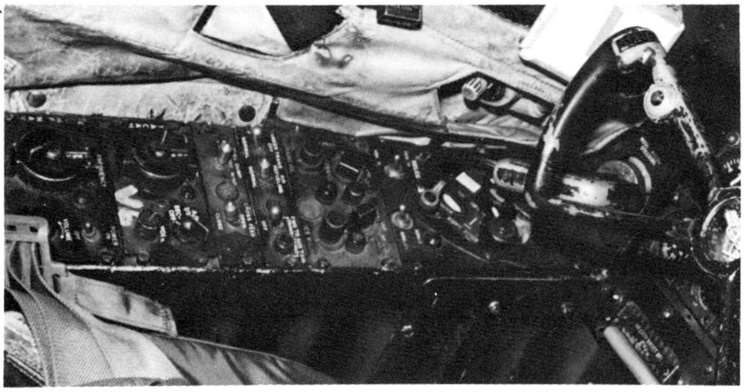

Bruce Trombecky

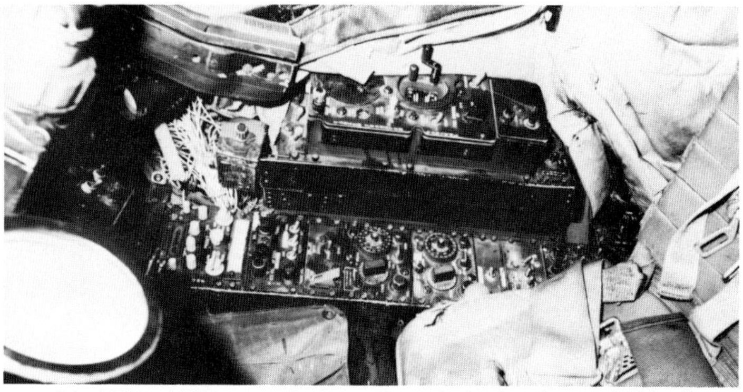

Black-painted KA-3B left and right forward cockpit side consoles. These accommodated most of the VHF and UHF communications radios and related equipment. Additionally, master crew communications radio panels also were located here. Vintage environmental insulation technology is apparent on cabin walls behind side consoles. Noteworthy, too, is the vintage "pepper grinder" com system tuner visible above the main right side panel.

COCKPIT EQUIPMENT (BOMBARDIER)

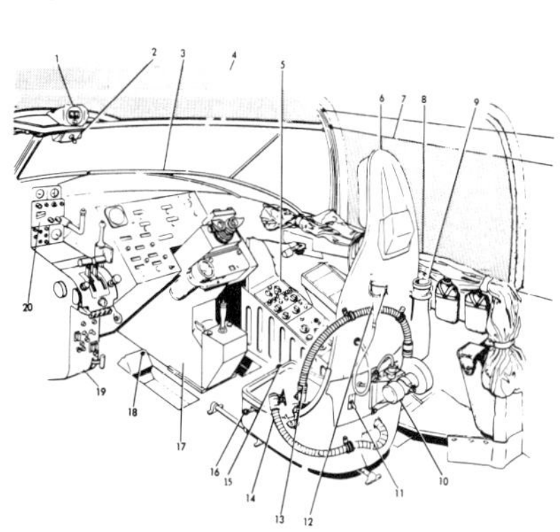

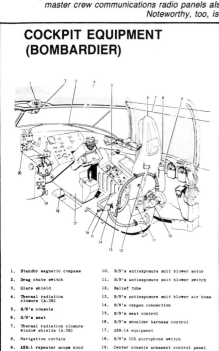

1. Standby magnetic compass
2. Drag chute switch
3. Glare shield
4. Thermal radiation closure (A-3B)
5. B/N's console
6. B/N's seat
7. Thermal radiation closure window shields (A-3B)
8. Navigation curtain
9. ASB-1 repeater scope hood
10. B/N's antiexposure suit blower motor
11. B/N's antiexposure suit blower switch
12. Relief tube
13. B/N's antiexposure suit blower air hose
14. B/N's oxygen connection
15. B/N's seat control
16. B/N's shoulder harness control
17. ASB-1A equipment
18. B/N's ICS microphone switch
19. Center console armament control panel
20. ATM monitor panel

COCKPIT EQUIPMENT (NAVIGATOR)

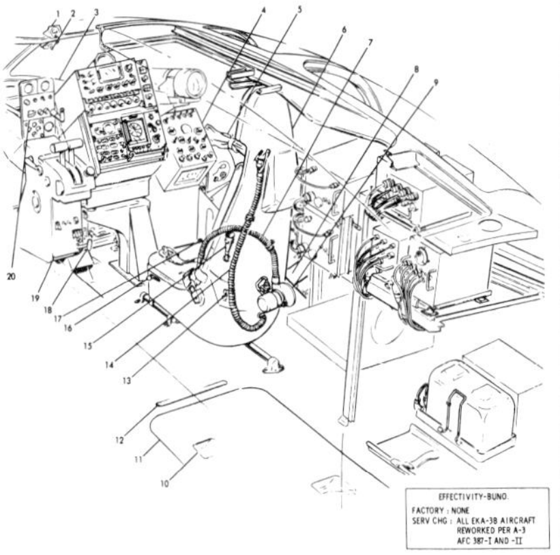

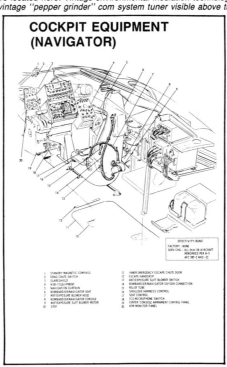

1 STANDBY MAGNETIC COMPASS
2 DRAG CHUTE SWITCH
3 GLARESHIELD
4 ASB-7 EQUIPMENT
5 NAVIGATION CURTAIN
6 BOMBARDIER/NAVIGATOR SEAT
7 ANTIEXPOSURE BLOWER HOSE
8 BOMBARDIER/NAVIGATOR CONSOLE
9 ANTIEXPOSURE SUIT BLOWER MOTOR
10 STEP
11 INNER EMERGENCY ESCAPE CHUTE DOOR
12 ESCAPE HANDGRIP
13 ANTIEXPOSURE SUIT BLOWER SWITCH
14 BOMBARDIER/NAVIGATOR OXYGEN CONNECTION
15 RELIEF TUBE
16 SHOULDER HARNESS CONTROL
17 SEAT CONTROL
18 ICS MICROPHONE SWITCH
19 CENTER CONSOLE ARMAMENT CONTROL PANEL
20 ATM MONITOR PANEL

COCKPIT EQUIPMENT (REAR)

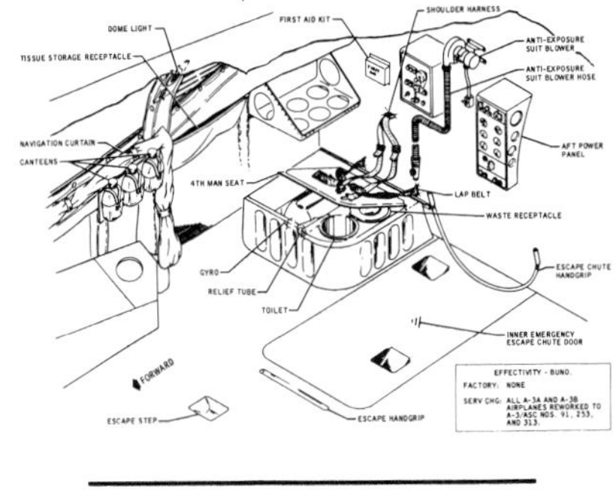

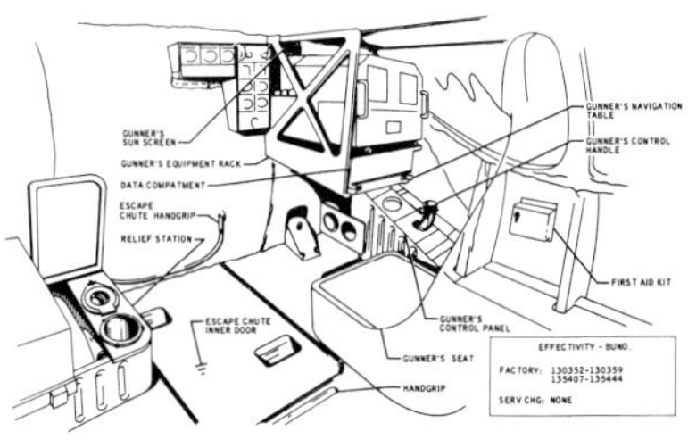

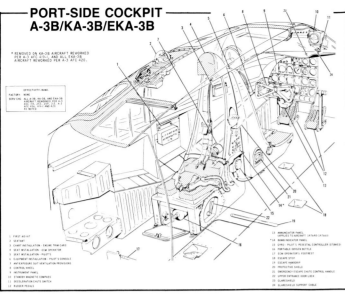

PORT-SIDE COCKPIT
A-3B/KA-3B/EKA-3B

* REMOVED ON KA-3B AIRCRAFT REWORKED PER A-3 AFC 419-1, AND ALL EKA-3B AIRCRAFT REWORKED PER A-3 AFC 420.

1 FIRST AID KIT
2 SEAT BELT
3 CHART INSTALLATION - ENGINE TRIM CARD
4 SEAT INSTALLATION - ECM OPERATOR
5 SEAT INSTALLATION - PILOT'S
6 EQUIPMENT INSTALLATION - PILOT'S CONSOLE
7 ANTIEXPOSURE SUIT VENTILATION PROVISIONS
8 CONTROL WHEEL
9 INSTRUMENT PANEL
10 STANDBY MAGNETIC COMPASS
11 DECELERATION CHUTE SWITCH
12 RUDDER PEDALS
13 ANNUNCIATOR PANEL
14 BOMB INDICATOR PANEL
15 GYRO - PILOT'S PEDESTAL CONTROLLER (STOWED)
16 PORTABLE OXYGEN BOTTLE
17 ECM OPERATOR'S FOOTREST
18 ESCAPE STEP
19 ESCAPE HANDGRIP
20 PROTECTIVE SHIELD
21 EMERGENCY ESCAPE CHUTE CONTROL HANDLE
22 UPPER ENTRANCE DOOR LOCK
23 GLARESHIELD
24 GLARESHIELD SUPPORT CABLE

The crew entrance tunnel is equipped with an in-cabin hatch (open in this view) that serves as flooring when closed. The tunnel walls are equipped with handgrips and foot slots to facilitate ingress and egress.

ERA-3B main instrument panel is painted black to reduce glare in sunlight. Instrumentation is standard for type, with most flight-related instruments clustered on the panel in front of the pilot, on the left. Powerplant instrumentation is surprisingly limited. The throttle quadrant is mounted on the center console. Navigation and Bendix-manufactured radar equipment (which includes the small CRT) is visible on the right side of the panel.

Old gunner equipment console and aft-facing seating position (port side) in ERA-3B now is occupied by systems-related electronic indicators and jump-seat location. Back-pack-type parachute also serves as seat-back padding.

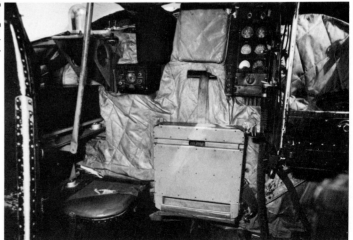

KA-3B cockpit view looking aft. An electrical system control panel is visible center upper right. Noteworthy is the insulation material attached to all exposed back bulkhead surfaces. The "honey bucket" is visible on the left.

View looking aft into the passenger cabin of BuNo 144857. This July, 1975, photograph provides insight into the arrangement of the five-seat VIP accommodation used by the Secretary of the Navy and the CNO.

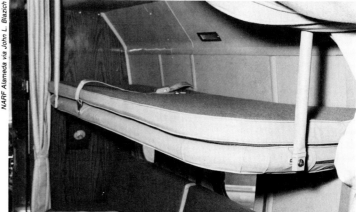

As shown on this view of the right side of the VIP cabin, BuNo 144857 could be fitted with two bunks. When not used as a bunk, the lower unit served as a side-facing bench.

This view looking forward into the cockpit of BuNo 144857, a VIP-configured TA-3B, was taken at NARF Alameda in December, 1973. The device on the table in the back of the pilot's seat is a pressure controller.

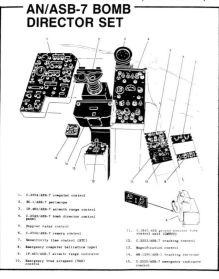

AN/ASB-7 BOMB DIRECTOR SET

1. C-2554/ASB-7 computer control
2. SU-1/ASB-7 periscope
3. IP-460/ASB-7 azimuth range control
4. C-2549/ASB-7 bomb director control panel
5. Doppler radar control
6. C-2550/ASB-7 camera control
7. Sensitivity time control (STC)
8. Emergency computer ballistics input
9. IP-463/ASB-1 azimuth range indicator
10. Emergency true airspeed (TAS) control
11. C-3645/ASB ground monitor tone control unit (GMTCU)
12. C-2553/ASB-7 tracking control
13. Magnification control
14. MX-1295/ASB-1 tracking recorder
15. C-2555/ASB-7 emergency indicator control

KA-3B aft facing station on the port side of the main cockpit cabin. Equipment includes a roof-suspended periscopic sextant and work table—the latter originally serving as a mounting frame for bombing system equipment. Attached to its sides are oxygen system hoses.

Aircraft oxygen is carried in ball-shaped containers that are easily removed from their under-fuselage compartment for refilling.

95

The electronic warfare compartment of the ERA-3B accommodates two crew members in an almost totally enclosed cabin. The area is accessed through the main cabin entry hatch, which is located immediately to the rear of the cockpit. Electronic warfare panels are mounted on the port side of the aircraft, with seating on the starboard side. The EW systems themselves have been constantly updated throughout the history of the EW-configured A-3 program.

ECM EVALUATOR'S CONSOLE ARRANGEMENT

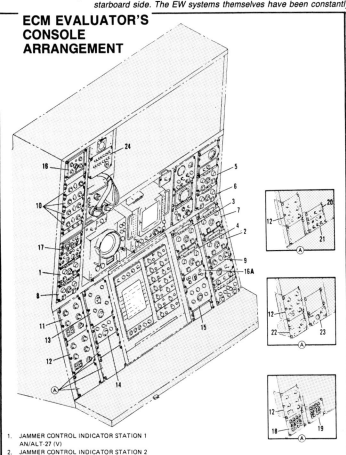

1. JAMMER CONTROL INDICATOR STATION 1 AN/ALT-27 (V)
2. JAMMER CONTROL INDICATOR STATION 2 AN/ALT-27 (V)
3. JAMMER CONTROL INDICATOR STATION 3 AN/ALT-27 (V)
4. JAMMER CONTROL INDICATOR STATION 4 AN/ALT-27 (V)
5. JAMMER CONTROL INDICATOR STATION 5 AN/ALT-27 (V)
6. JAMMER CONTROL INDICATOR STATION 6 AN/ALT-27 (V)
7. JAMMER CONTROL INDICATOR STATION 7 AN/ALT-27 (V)
8. JAMMER CONTROL INDICATOR STATION 8 AN/ALT-27 (V)
9. MASTER STEERING CONTROL AN/ALT-27
10. INDIVIDUAL ANTENNA STEERING CONTROL (FIXED) AN/ALT-27
11. JAMMER CONTROL INDICATOR (AFT) STATION 9 AN/ALT-32H
12. JAMMER CONTROL INDICATOR (FWD) STATION 10 AN/ALT-32H
13. BATWING SWITCH CONTROL AN/ALT-32H/ALR-43
14. DECM SYSTEM CONTROL (AFT) AN/ALQ-41
15. DECM SYSTEM CONTROL (FWD) AN/ALQ-41
16. CHAFF PROGRAMMER AN/ALE-29
16A. CHAFF CONTROL AN/ALE-29
17. CHAFF CONTROL UNIT AN/ALE-2
18. JAMMER CONTROL UNIT (PORT POD) STATIONS 11A, B, C and D AN/ALQ-76
19. JAMMER CONTROL UNIT (STARBOARD POD) STATIONS 12A, B, C AND D AN/ALQ-76

ECM EVALUATOR'S CONSOLE ARRANGEMENT

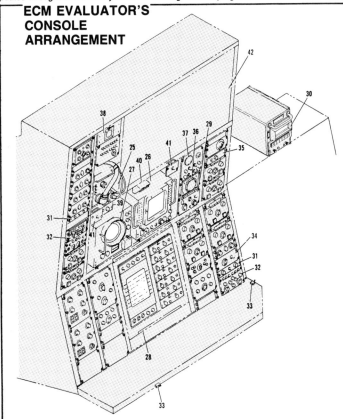

*20. LAUNCH PANEL AN/AQM-37A
*21. CONTROL UNIT AN/AQM-37A
*22. FORWARD FIRING CHAFF ROCKET CONTROL UNIT (PORT POD) AN/ALE-25
*23. FORWARD FIRING CHAFF ROCKET CONTROL UNIT (STARBOARD POD) AN/ALE-25
24. MASTER RADIATE CONTROL
25. PULSE ANALYZER POWER SUPPLY AN/ULA-2
26. PULSE ANALYZER AN/ULA-2
27. AZIMUTH INDICATOR IP-81A/APA-69
28. ESM RECEIVER AND CONTROL DISPLAY AN/ALR-43
29. DF CONTROL WJ/APA-69
30. FREQUENCY MEASURING UNIT AN/ALR-43
31. UHF TRANSFER PANEL (2)
32. ICS PANEL (2)
33. ICS SWITCH (2)
34. ECM CONSOLE LIGHT CONTROL PANEL
35. BDHI SELECTOR SWITCH CONTROL PANEL ASN-50/ASN-41
36. BDHI
37. RAT POWER CONTROL PANEL
38. TEMPERATURE INDICATOR FORWARD EQUIPMENT COMPARTMENT
39. LOW CABIN PRESSURE WARNING LIGHT
40. CLOCK
41. ECM EVALUATOR OXYGEN CONTROL (2)
42. ECM CIRCUIT BREAKER PANEL NO. 120

*ALTERNATE EQUIPMENT-MISSION DETERMINES CONFIGURATION.

FORWARD ECM RACK AND EQUIPMENT (ERA-3B)

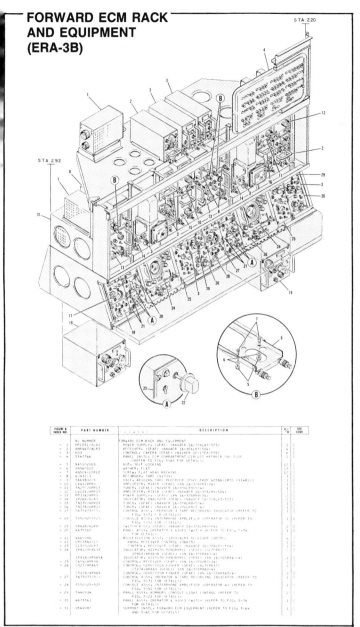

AFT ECM RACK AND EQUIPMENT (ERA-3B)

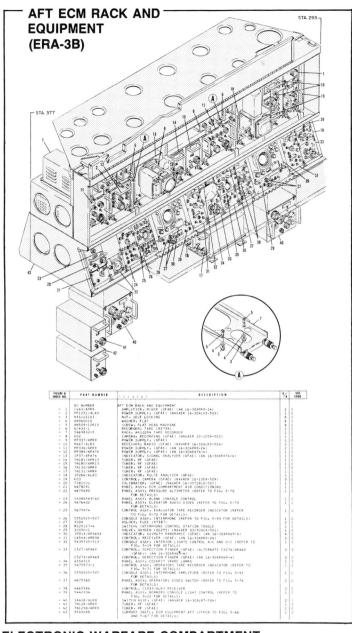

AR-200 SIGNAL DATA WIDEBAND RECORDER AND ELECTRONIC WARFARE COMPARTMENT (EA-3B)

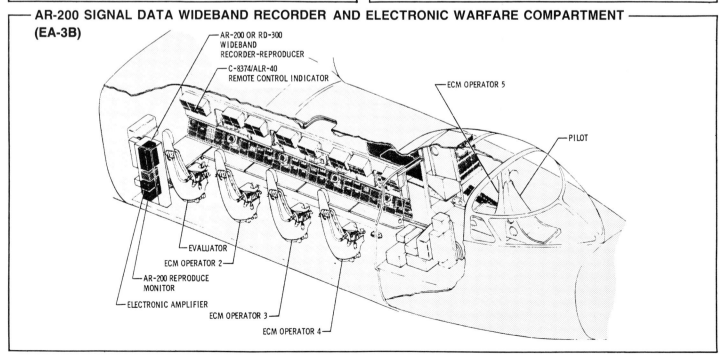

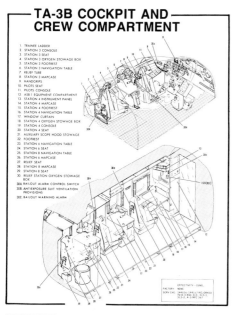
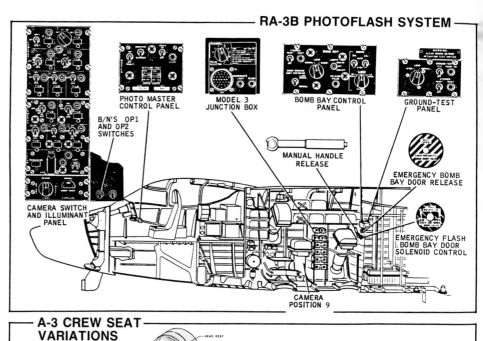

A Block 4 Hughes AWG-9 system for the Grumman F-14A as installed in the Hughes TA-3B testbed aircraft. The radar hand control is mounted to the right.

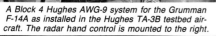

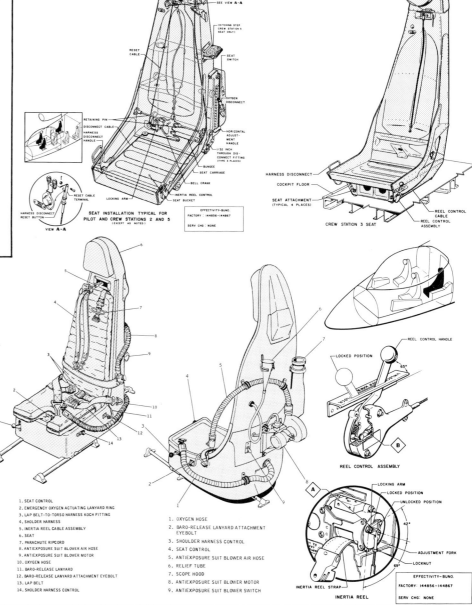

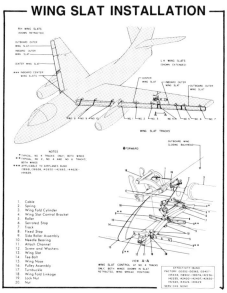
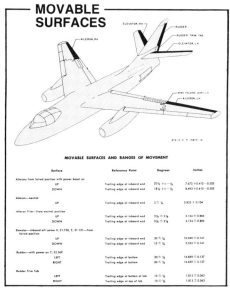
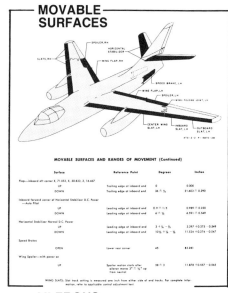
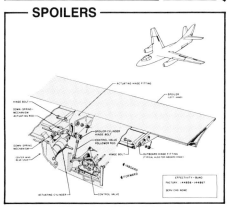
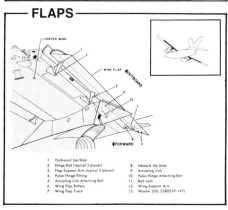
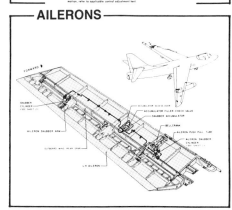
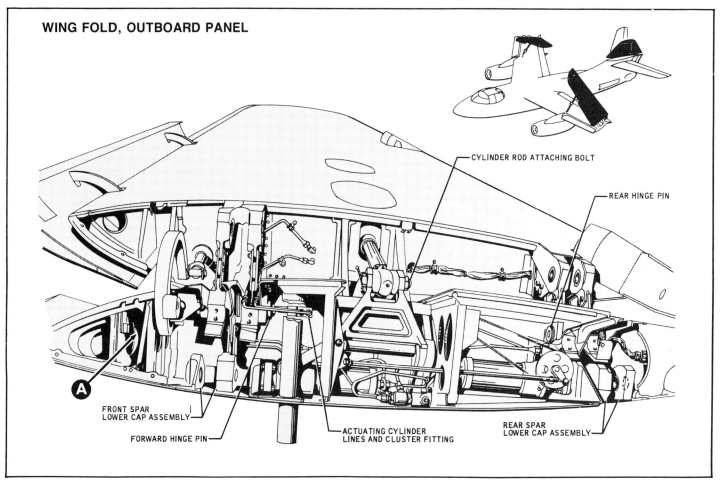

Wing fold on the A-3 was necessitated by the Navy's stringent carrier compatibility requirements. Accordingly, Douglas designed the aircraft so that both outer wing panels and the vertical tail could be folded at will. Hydraulic actuators provide the power required to fold and unfold the panels while sliding pins lock the wings in place.

Topside view of folded outer wing panel indicates tight tolerances at the fold point edge assemblies. Ailerons are deactivated during folding.

The A-3 is equipped with a conventional, single-piece flap on the inboard trailing edge of each wing. The flaps are capable of being utilized throughout most of the aircraft's flight envelope and are hydraulically actuated. Downward travel of the flap is 36° plus or minus 1/2°.

Each wing is equipped with aerodynamically actuated leading edge slats. These are extended automatically as a function of airspeed and angle-of-attack.

In order to preserve the relatively good drag characteristics of the A-3, every effort within the constraints of cost was made to keep the aircraft aerodynamically clean. One example of this was the fairly complex engine nacelle pylon fairing assembly and its attached flap segment. Refined geometries permitted the thin and unsupported rear segment to rotate downward with the flap while maintaining complete mechanical integrity.

FIN AND TIP INSTALLATION

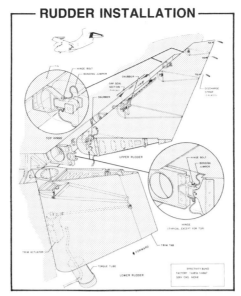

RUDDER INSTALLATION

HORIZONTAL STABILIZER INSTALLATION

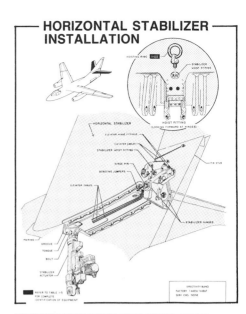

The vertical fin, like the wing outer panels, is foldable to accommodate aircraft carrier deck clearance requirements. The fold is to the starboard side of the aircraft and can be accomplished from the cockpit. The vertical fin is hinged at two points and once a release pin is hydraulically removed, folding of the surface is hydraulically accomplished. A single actuator, mounted in the lower fin segment, imparts the folding action.

The A-3's vertical fin tips have accommodated many sensor and tail warning systems throughout the aircraft's operational and flight test histories. Three tips, including (from l. to r.) that for a VA-3B, an ERA-3B, and a TA-3B, are shown. The ERA-3B tip is the only one of these three still to contain a functioning warning system. The small fairing seen on the tail of the TA-3B originally was designed to accommodate a tail warning sensor but now is thought to be empty.

The A-3, because of the variety of operational and test missions it has been required to accommodate over the course of its life span, has gone through many configuration changes. The most visible of these has affected its empennage design and three of the more noteworthy configurations are illustrated in these photos. Empennages shown (from l. to r.) include the aerodynamic dummy tail of an A3D-1, the tail of an A3D-2T, and the electronics compartment-configured TA-3B.

Three more noteworthy empennage sections are illustrated in these photos of (from l. to r.) a mid-life version of the ERA-3B tail, a late-version of the ERA-3B tail, and the experimental missile control system antenna fairing of NA-3A, BuNo 135409. The ERA-3B, because of the nature of its mission, has utilized empennage section space to accommodate several different types of automatic chaff dispensing equipment. Chaff is ejected through the rectangular holes.

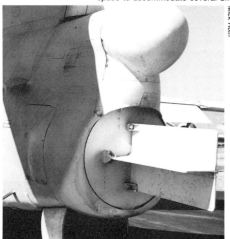

A variety of electronic warfare antenna configurations have been seen mounted in the A-3's tail. Some have been faired and some have not.

The A-3 is equipped with two single-piece, all-metal, hydraulically actuated airbrakes, with one mounted on each side of the aft fuselage, several feet to the rear of the wing trailing edge. A single hydraulic actuator opens and closes each airbrake upon pilot command. The airbrake, within limits, can be utilized throughout much of the A-3's flight envelope and are particularly effective as landing aids when rapid deceleration is required to destroy wing lift.

SPEED BRAKE AND CONTROL

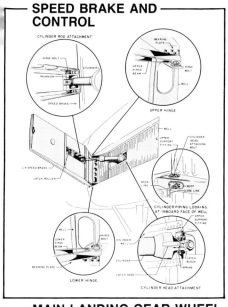

NOSE LANDING GEAR INSTALLATION

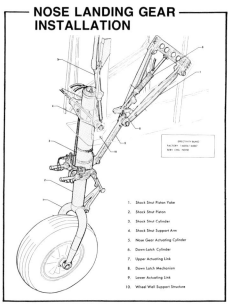

1. Shock Strut Piston Yoke
2. Shock Strut Piston
3. Shock Strut Cylinder
4. Shock Strut Support Arm
5. Nose Gear Actuating Cylinder
6. Down-Latch Cylinder
7. Upper Actuating Link
8. Down Latch Mechanism
9. Lower Actuating Link
10. Wheel Well Support Structure

NOSE LANDING GEAR INSTALLATION

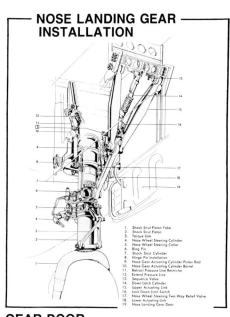

1. Shock Strut Piston Yoke
2. Shock Strut Piston
3. Torque Link
4. Nose Wheel Steering Cylinder
5. Nose Wheel Steering Collar
6. Ring Pin
7. Shock Strut Cylinder
8. Hinge Pin Installation
9. Nose Gear Actuating Cylinder Piston Rod
10. Nose Gear Actuating Cylinder Barrel
11. Retract Pressure Line Restrictor
12. Extend Pressure Line
13. Sequence Valve
14. Down Latch Cylinder
15. Upper Actuating Link
16. Lock Down Limit Switch
17. Nose Wheel Steering Two Way Relief Valve
18. Lower Actuating Link
19. Nose Landing Gear Door

MAIN LANDING GEAR WHEEL AND BRAKE INSTALLATION

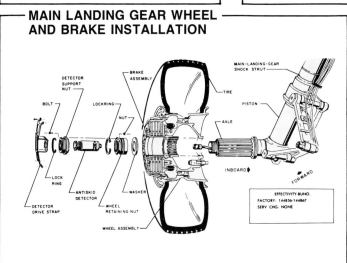

MAIN LANDING GEAR DOOR INSTALLATION

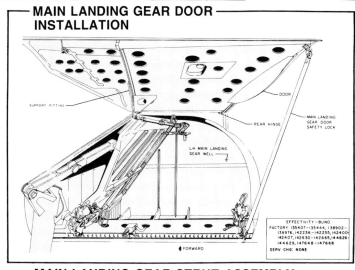

MAIN LANDING GEAR INSTALLATION

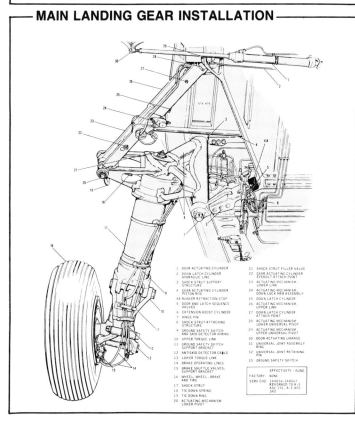

1. Door Actuating Cylinder
2. Down Latch Cylinder Hydraulic Line
3. Shock Strut Support Structure
4. Gear Actuating Cylinder Piston Rod
4A. Rubber Retraction Stop
5. Door and Latch Sequence Valves
6. Extension Boost Cylinder
7. Hinge Pin
8. Shock Strut Attaching Structure
9. Ground Safety Switch and Skid Detector Wiring
10. Upper Torque Link
11. Ground Safety Switch Support Bracket
12. Antiskid Detector Cable
13. Lower Torque Link
14. Brake Operating Lines
15. Brake Shuttle Valves Support Structure
16. Wheel, Wheel, Brake and Tire
17. Shock Strut
18. Tie Down Spring
19. Tie Down Ring
20. Actuating Mechanism Lower Pivot
21. Shock Strut Filler Valve
22. Gear Actuating Cylinder Eyebolt Attach Point
23. Actuating Mechanism Lower Link
24. Actuating Mechanism Lower Link Assembly
25. Down Latch Arm Assembly
26. Actuating Mechanism Upper Link
27. Down Latch Cylinder Attach Point
28. Actuating Mechanism Lower Universal Pivot
29. Actuating Mechanism Upper Universal Pivot
30. Door Actuating Linkage
31. Universal Joint Assembly
32. Universal Joint Retaining Pin
33. Ground Safety Switch

MAIN LANDING GEAR STRUT ASSEMBLY

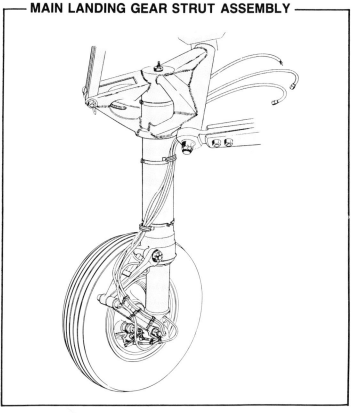

The nose landing gear assembly is fairly conventional and designed to be relatively lightweight and simple. A single axle and conventional fork arrangement support the wheel and tire assembly. Retraction is hydraulically actuated and gear strut movement is forward. Steering is hydraulically controlled by a pair of small actuators. A single drag link prevents over-extension to the rear. The nose gear well is covered by a single, dual-hinged door.

The nose gear anti-torque scissor assembly, which also serves to impart steering moments to the wheel and tire assembly, usually is disconnected when it becomes necessary to ground tow the A-3. The gear well accommodates the hydraulic actuators required to retract and extend the nose landing gear as well as the gear itself in the retracted condition. A small taxi light, mounted below the LSO approach light box, is attached to the nose gear door.

The main landing gear assemblies are of simple but robust construction. Their dynamic loads are passed off to the fuselage structure through simple hinged mounts. Retraction of each main gear assembly is facilitated by a single, large hydraulic actuator mounted transversely and pulling the gear inward, into the gear well. Each gear well door is sequenced to close via a single hydraulic actuator, after the gear is retracted.

The main gear well is relatively uncluttered and is capacious enough to accommodate the A-3's fairly large tire and wheel assembly. At the bottom of the starboard gear well is the aft tank pressure fueling attachment. A small keel-like bulkhead is basically all that divides the space occupied by the two wells.

The anti-torque link assembly for the main gear is mounted aft of the main gear strut assembly. It also serves to support brake system hydraulic lines.

The A-3's main gear strut and axle assembly are extraordinarily robust to accommodate the loads imparted during carrier landings and takeoffs. At least two major brake configurations have been utilized by the A-3 during the course of its production history. One, shown, resulted in the actual brake assembly being mounted external to the wheel; the other was more conventional in appearance, with the brake being accommodated inside the wheel assembly.

Because of the A-3's rotational requirements during takeoff and landing and the fixed geometric dictates of its main landing gear assembly, the aircraft is equipped with a tail wheel and strut that serve primarily to protect the empennage section and aft lower fuselage from damage. The wheel and its associated assembly are hydraulically extended and retracted, usually in concert with main landing gear extension and retraction.

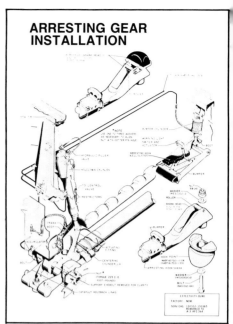

ARRESTING GEAR INSTALLATION

A conventional tail hook is attached to the A-3, just aft of its tailwheel assembly. Extension, which is accomplished hydraulically, is upon pilot command. During flight, the tailhook remains exposed, but recessed into a cavity on the empennage bottom. A-3 tailhook colors usually are black and white.

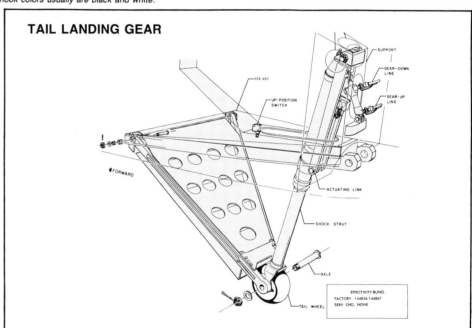

TAIL LANDING GEAR

A small bumper is recessed into the fuselage bottom cavity to soften the upward movement of the tailhook once a carrier trap has been accomplished.

The A-3 is equipped with a 24 feet-in-diameter ring-slot type nylon drag chute that is housed in a compartment on the underside of the extreme aft section of the fuselage empennage. When used in conjunction with the A-3's very effective main gear brakes and massive airbrakes, deceleration is extremely rapid. Though its use is pilot-discretionary, the drag chute is not used during carrier operations; it is, however, deployed when necessary during land-based assignments.

CATAPULT HOOK

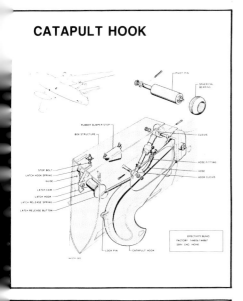

Retractable catapult hooks on the underside of the forward fuselage attach the catapult bridle to the aircraft and impart the energy generated by the steam actuated system to the aircraft, thus permitting rapid acceleration to takeoff speeds. The bridle often is cast into the sea and thus is considered to be expendable.

AN/ASB-12 RADAR INSTALLATION

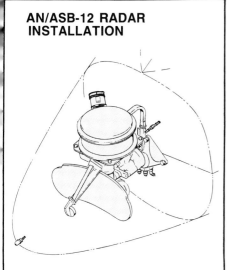

A-3 EW SYSTEMS/EQUIPMENT

A-3A:					
EA-3A:	APR-9	ALQ-2	**EKA-3B:**	ALQ-41	ALT-27
	APR-13	ALQ-23		ALQ-51	ALA-3
	APA-69	APA-74		ALQ-92	ULA-2
	ALA-3			ALR-28	C-7317/AL DF Display
	ALR-3			ALR-29	
A-3B:	ALQ-19	ALQ-100 (?)		ALR-30	
	ALQ-32	ALE-2		APR-32	
	ALQ-35	ALT-27 (?)		ALE-1/ALE-2 (combination of ALE-1 & ALE-2 components)	
	ALQ-41				
	ALQ-51		**RA-3B:**	ALQ-35	APR-25
	ALQ-55			ALQ-41	APR-27
EA-3B:	ALA-3	ALQ-55		ALQ-51	APR-30 (?)
	ALR-3	ALQ-100 (?)		ALQ-55	
	APA-69	ALR-40		ALQ-100 (?)	
	APA-74			ALE-29	
	APR-9				
	APR-13		**ERA-3B:**	ALQ-41	ALE-25 pod
	ALQ-35			ALQ-31 pod	WJ-8570 system
	ALQ-41			ALQ-76 pod	MD-492/ULT modulator
	ALQ-51			ALQ-167 pod (formerly DLQ-3)	APA-69
KA-3B:	ALQ-41			ALT-27	ALR-43
	ALQ-51			ALT-32	ALE-29
	ALQ-55			ALE-2	ULA-2
	ALQ-35		**TA-3B**		

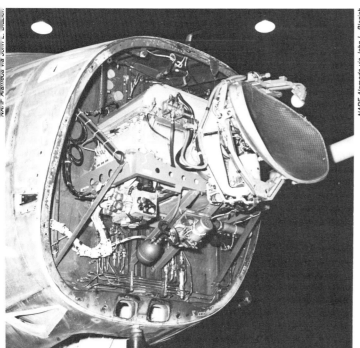

Two views of the TA-3B's AN/ASB-12 radar system showing the RT/Syn and vacuum pump assemblies as fitted to BuNo 144858 of RVAH-3 at NAS Key West, Florida. This aircraft was used to train navigators assigned to RVAH squadrons flying the RA-5C. The radar's parabolic dish (which was articulated in both azimuth and elevation) and related mechanical components were fairly standard technology for their day.

RADIO AND RADAR EQUIPMENT AND ANTENNAS

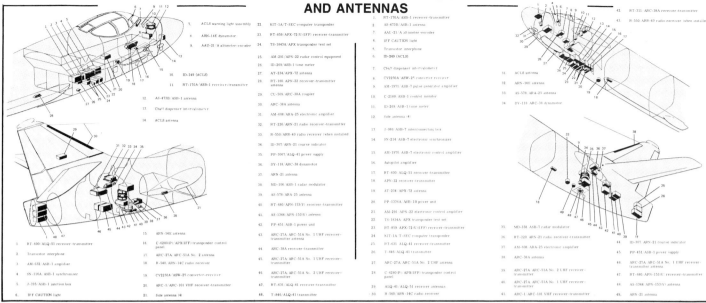

The ERA-3B is equipped with four externally-mounted ram-air turbines (RATs). As the A-3's forward velocity in flight is sufficient to turn each RAT's small, variable-pitch propeller at a very high rate of speed, significant electrical power can be generated and utilized to power on-board electronic warfare emitters and related equipment. Other protuberances visible include antenna, cooling intakes and related dump ports for electrical equipment, and the ventral antenna fairing canoe.

The ERA-3B's primarily plexiglas ventral canoe fairing covers an extensive array of transmitting and receiving antenna optimized for the aircraft electronic warfare jamming mission. Power for the system is generated by the A-3's J57 engines and the four externally-mounted RATs.

Cont. from p.91

retracted by 3000-psi utility hydraulic system pressure. The arresting hook holddown unit consists of an air chamber containing 675- to 700-psi air pressure. The air gage, which also supplies air pressure for extending the hook, is located aft, on the external left-hand side of the fuselage. The hook is mechanically latched in the up position.

The arresting hook handle, which is hook-shaped for easy identification, is located on the left-hand side of the center console. Movement of the handle to the hook-down position mechanically releases the hook uplatch and actuates the solenoid-operated hydraulic selector valve to relieve hydraulic pressure in the actuating piston, thereby allowing the hook to extend by air pressure in the holddown unit. Emergency operation of the arresting hook is accomplished by pulling the hook cable in the bomb bay.

Cont. on p.110

The RATs, because of their minimal cross-sectional area and uncluttered mounting positions, create only minimal drag at the ERA-3B's nominal cruising speed and altitude. RAT propeller speed can be regulated by controlling the propeller blade pitch angles. This, in turn, permits exacting control of the energy being generated by each RAT in conjunction with aircraft airspeed, and electronic warfare equipment energy requirements.

LORAN and COMINT/SIGINT antennas were utilized on some A-3 configurations, the antenna farms associated with these varying in size and configuration.

Many A-3s have been configured to carry ordnance on underwing pylons. The pylons are removable, and are mounted outboard of the engine nacelles.

An AN/AST-4 pod is representative of just one of the many external loads that can be attached to the A-3's wing pylons.

Fifteen or more antenna associated with electronic warfare jamming systems can be housed in the A-3's fiberglass ventral canoe. Several ventral canoe configurations have been utilized during the course of the A-3's EW history.

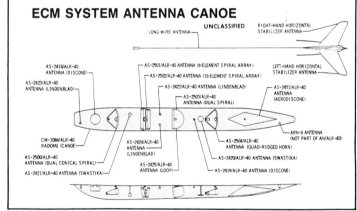

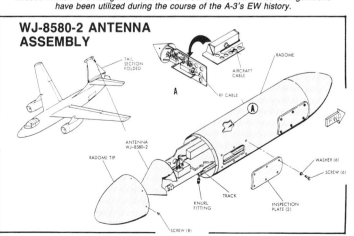

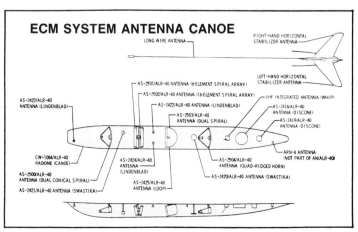

109

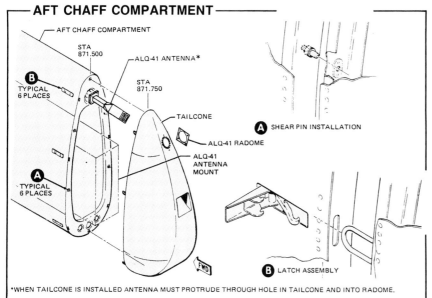

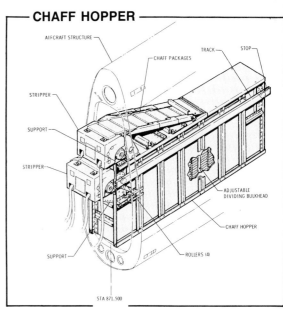

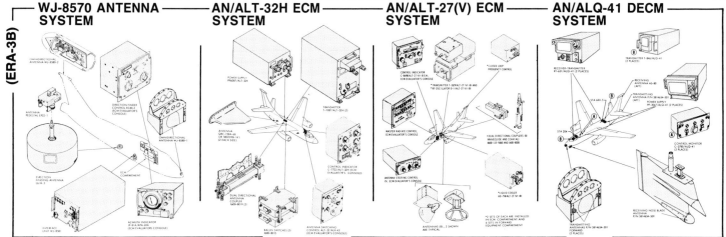

Avionics and Communications—The KA-3B principal navigation system consists of an airborne AN/ASB-7 or AN/ASB-1A radar. A more modern color radar is to replace it in the near future. The AN/APN-153 doppler allows for continuous dead reckoning (DR) and for continuous readout of groundspeed and drift angle information. Other navigation systems include the AN/ARA-5 UHF direction finder, the AN/ARN-14E VOR homing set, the AN/ARN-118 TACAN, the AN/APN-22 radio altimeter, the AN/AQU-1A sextant, and LORAN C. Some aircraft have also been retrofitted with an *Omega* system.

IFF and altitude reporting is provided by AN/APX-72. A Sperry S-5 autopilot is fitted and the AN/ASW-25 automatic carrier landing system has been added by Aircraft Service Change No. 467.

Principal communications systems are the AN/ARC-84 HF, the AN/ARC-51A UHF, and the AN/ARC-101 VHF radio sets.

No less than three AN/ALQ-167 Multiple Environment Threat Emitters (METEs) are seen attached to an A-3 port outer wing pylon. These units, which contain an AN/DLQ-3 transmitting unit as their primary power source, usually have four antenna resembling missile fins, at their aft end. Primary color usually is medium blue, with nose caps having a variety of colors. The transmitting antennas seen protruding from the nose caps usually have two different shapes: bulbous and short and blunt.

Nose of TA-3B illustrates variation in cockpit canopy framing, lack of external antenna and other modifications. Nose cone opens vertically for internal systems access. Pitot is mounted atop nose, just ahead of windscreen.

Early ERA-3B nose illustrates variation in cockpit canopy framing, various electronic warfare antenna positions, the nose-mounted pitot, the location of the pilot's side windshield wiper blades, and the forward mounts for the inflight refueling boom.

Starboard side of late ERA-3B nose illustrates variation in cockpit canopy framing, various electronic warfare antenna positions, the nose-mounted pitot installations, location of the forward RAT, and the nose optical ports.

Port side of late ERA-3B nose illustrates variation in cockpit canopy framing, various electronic warfare antenna locations, the nose-mounted pitot installations, location of the forward RAT, and the mounting points for the inflight refueling boom.

Starboard side of EKA-3B nose illustrates variation in cockpit canopy framing, antenna fairing under nose, small antenna under nose, pitot location ahead of windscreen external bracing, and hot air dump aft of numbers.

Port side of A-3B nose illustrates variation in cockpit canopy framing, pitot location ahead of windscreen, access panels, antenna fairing on nose gear well door, EW antenna under nose, and small NACA intake ahead of access panel.

Back from a war cruise aboard the USS "Coral Sea" (CVA-43), during which the aircraft flew sixteen bombing missions, A-3B, BuNo 147655, of VAH-2 was photographed at Long Beach during 1965.

A-3 electronic warfare configurations vary greatly. Port side of EA-3B nose illustrates sensor antenna fairing and side-mounted comint antennas. Original A-3 nose configuration is apparent. Noteworthy are hot air dump ports.

BuNo 144867, the TA-3B which was bailed to Hughes Aircraft Company for use as a radar testbed, was photographed on arrival at the Hughes plant on October 28, 1968. The aircraft is now used for testing the F-14D radar system.

Soon after obtaining BuNo 144867, Hughes fitted it with sensors and recording equipment. Note that the refueling probe has been replaced by a sensor probe for test data measuring and recording.

Various missile control and homing systems have been tested while mounted in the A-3's fairly capacious nose. Modification illustrated is for one of several cruise missiles being explored for Navy use in ship-launched form.

A unidentified missile nose, possibly a "Tomahawk", is seen protruding from BuNo 138938 of the Point Mugu Test Center. A wave-guide fairing can be seen coursing the entire length of the A-3's starboard fuselage.

Another unidentified nose modification, seen protruding from NMC BuNo 142630. If the modification is the forward component of a missile, it is a new type not previously identified. The flat plate area of the nose where the missile is attached is noteworthy.

BuNo 142256 was photographed on the ramp at NAS Alameda in June, 1969, after personnel from the Naval Air Rework Facility Alameda had installed a container on the fuselage side in preparation for the use of the aircraft as a testbed for sonobuoys.

Bulbous nose of NRA-3B, BuNo 144825. Though appearing to increase the cross sectional drag of the aircraft, the nose is actually quite efficient as it falls well within the fineness ratio parameters of the basic aircraft fuselage.

Still another nose modification involving a seeker head and an optical tracking unit (possibly similar to the Northrop TCS found on the Grumman F-14A). Flat plate area of nose apparently poses no aerodynamic problem at relatively low airspeeds.

Taken at NAS Point Mugu in 1970, this line-up of NMC test aircraft shows a variety of nose configurations and other modifications. Side number 70 is an NA-3A (BuNo 135418) which was damaged in 1974 and is now on display at the Naval Air Museum at NAS Pensacola; side number 75 is the well-known NRA-3B bearing BuNo 144825 and is still in use at Point Mugu; and side number 71, an NRA-3B (BuNo 142667), and side number 74, an NA-3B (BuNo 138938), are both current.

Hughes AN/AWG-9 radar designed for the General Dynamics/Grumman F-111B was flight tested aboard a Hughes-operated TA-3B, BuNo 144867. Numerous modifications were required to accommodate the test equipment and the weight of the radar.

TA-3B, BuNo 144867, with F-111B nose radome in place and i.r. seeker assembly slung independently underneath. Visible under the wing root section is part of the scabbed on wire ducting, which traveled most of the way to the aircraft's tail.

A-3B, assigned to the Naval Missile center and equipped with a variety of electronic warfare systems, is seen mounting a pair of AN/ALD-6 (XN-1) passive receiving antennas, similar to those now found on some U-2 variants.

Fitted with ventral and tail radomes and carrying an unidentifed jammer pod beneath its left wing, this NRA-3B of the Pacific Missile Test Center was photographed at Scott AFB, Illinois, on March 5, 1972.

Modifications required to accommodate the F-111B's radar system were substantial and included the installation of a liquid cooling system in the extreme aft end of the aircraft. Visible on the starboard side are wave guide fairings.

View of the port side of BuNo 144867 showing the additional wave guide fairings, the APU intake and the port side cryogenic cooling system access door. A dome fairing covered the aft end. Wave guides were run along the outside for simplicity.

RA-3B BOMB BAY

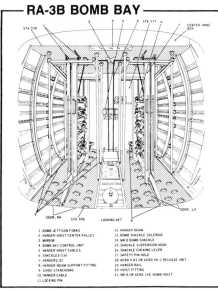

RA-3B PHOTOFLASH CARTRIDGES

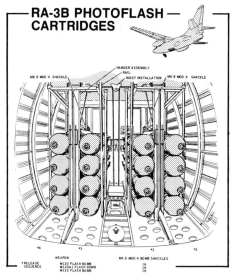

RA-3B PHOTOFLASH BOMBS

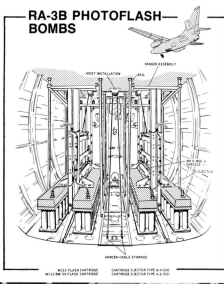

BOMB BAY COMPONENTS

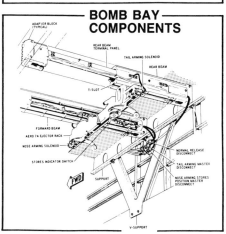

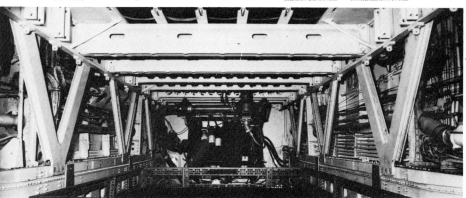

The weapons bay of all still-operational A-3s have, with only few exceptions, been reconfigured to accommodate fuel tanks, electronic warfare systems and/or compartments, and refueling system equipment. A KA-3B bay is shown, illustrating the internal structural modification required for this configuration.

BOMB BAY SPOILER

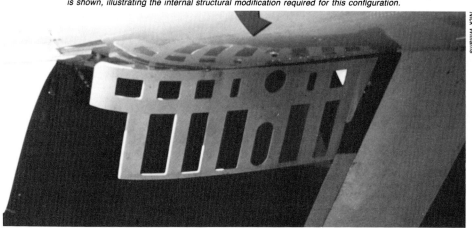

To reduce internal buffeting with the bay doors open, the A-3 was equipped with a bomb bay spoiler assembly that consisted of a heavily perforated door hinged at its forward end and moved into position hydraulically as the bay doors opened.

WEAPON LOADS

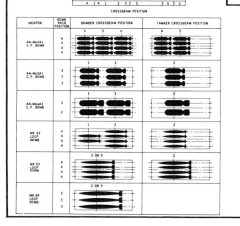

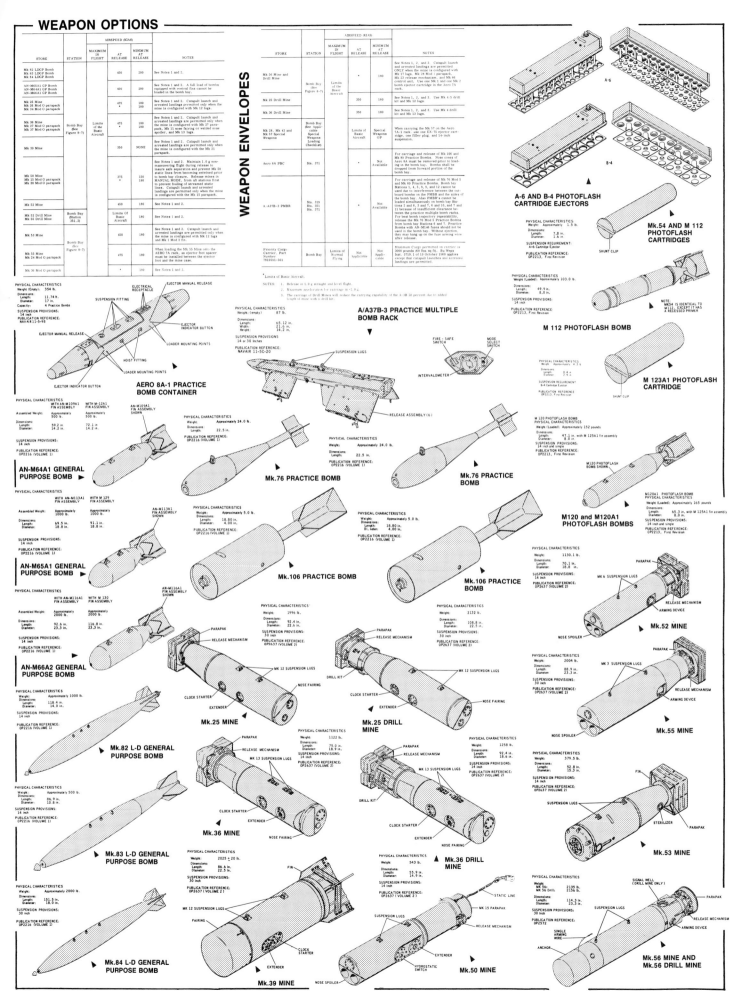

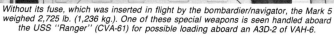

With its 61 in. (1.55 m) diameter and 8,500 lb. (3,856 kg.) weight, the Mark 6 was the widest and heaviest special weapon that could be carried by the A3D in its originally intended role as a carrier-based strategic bomber.

Without its fuse, which was inserted in flight by the bombardier/navigator, the Mark 5 weighed 2,725 lb. (1,236 kg.). One of these special weapons is seen handled aboard the USS "Ranger" (CVA-61) for possible loading aboard an A3D-2 of VAH-6.

The Mark 91 was one of the slimmer special weapons developed while the "Skywarrior" was being built. This Mark 91 empty case was photographed at the National Atomic Museum in Albuquerque.

Also photographed at the National Atomic Museum, this bomb case illustrates the external appearance of the Mark 12 special weapon. It had a length of 144 in. (3.66 m), a diameter of 22 in. (0.56 m) and a weight of 1,100 lb. (499 kg.).

With its 182 in. (4.62 m) length, the Mark 7 was the longest special weapon which the "Skywarrior" was cleared to carry and deliver. It was the first U.S. tactical nuclear weapon.

Rarely seen side-mounted forward fuselage pylon modification on BuNo 142667. The pylon carries an AN/ALQ-167 Multiple Environment Threat Emitter pod. Also visible is the forward section of an external wave guide fairing.

In this photograph taken at NAS Agana on October 9, 1962, a typical crew—pilot, photo-navigator/assistant pilot, and photo-technician/gunner—poses in front of an RA-3B from VAP-61 with an assortment of camera systems carried by this "Skywarrior" version. The photo technician could change magazines, adjust apertures, correct malfunctions, and rearrange cameras by accessing the pressurized camera compartment in flight.

LAU-24/A LAUNCHER AND ADU-238/A ADAPTER

In conjunction with the F-111B program, the Hughes TA-3B was modified so that the starboard outer wing pylon could accommodate the launch rail and pylon assembly needed for the Raytheon AIM-7 "Sparrow" air-to-air missile. Additionally, onboard avionics and the AN/AWG-9 radar were configured to provide AIM-7 interface requirements.

Wing mounted pylon assembly of the A-3 can be modified for test purposes to carry the AIM-9 "Sidewinder" air-to-air missile. An AIM-9H is shown with i.r. seeker head in place. Because of its size and related ability to carry extensive test instrumentation, the A-3 is ideal for this type of flight test work.

As predicted by Ed Heinemann, the twin gun turret (two 20mm M3L cannon with 500 rpg) soon proved of little use and was eventually replaced by a DECM system in a "dovetail" (or "duckbutt") fairing.

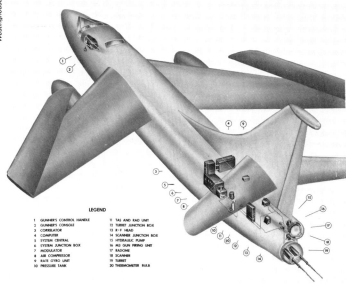

The location of the prinicpal components of the rear defense system initially fitted to most "Skywarriors" is illustrated by this artist rendering. Vibrations rendered the guns difficult to sight.

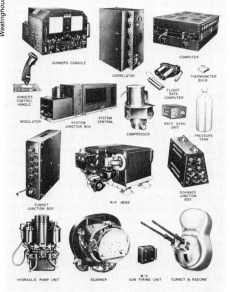

This composite photograph illustrates the principal components of the Aero 21B tail turret and its control mechanism.

TAILGUN INSTALLATION

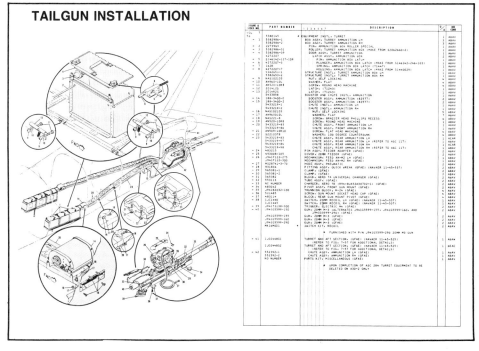

117

Specifications and Performance Data

Primary source documents for the specifications and performance tables were the following *Standard Aircraft Characteristics* charts:

Aircraft	SAC chart dated
A3D-1	September 1, 1958
YA3D-1P	July 1, 1954
A3D-2	April 15, 1961
A3D-2P	April 15, 1961
A3D-2Q	April 15, 1961
A3D-2T	April 14, 1958
KA-3B	March 3, 1958
(data for tanker-configured A3D-2)	

In the case of the XA3D-1, for which a SAC chart could not be obtained, specifications and performance data were compiled from a series of official documents and other published information. While every effort has been made to verify these data, they cannot be regarded as reliable as those extracted from SAC charts.

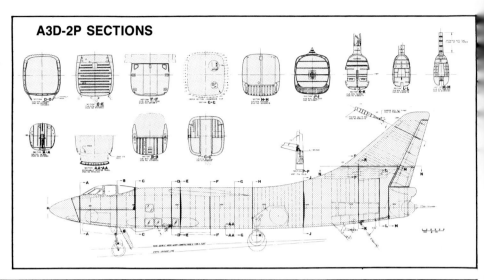

A3D-2P SECTIONS

DIMENSIONS, WEIGHTS AND TANK CAPACITY

	XA3D-1	A3D-1	YA3D-1P	A3D-2 (CLE wing)
DIMENSIONS:				
Span, ft	72.5	72.5	72.5	72.5
(m)	(22.10)	(22.10)	(22.10)	(22.10)
Span (wing folded), ft	48.2	49.2	48.2	49.4
(m)	(14.69)	(15.0)	(14.69)	(15.06)
Length/length with probe, ft	75.3/NA	74.4/NA	76.0/NA	74.7/78.2
(m)	(22.95/NA)	(22.68/NA)	(23.16/NA)	(22.77/23.84)
Height, ft	23.8	22.8	23.9	22.8
(m)	(7.25)	(6.95)	(7.28)	(6.95)
Height (tail folded), ft	23.35	15.1	16.5	15.9
(m)	(7.12)	(4.60)	(5.03)	(4.85)
Wing area, sq ft	779	779	779	812
(sq m)	(72.37)	(72.37)	(72.37)	(75.44)
WEIGHTS AND LOADINGS:				
Empty weight, lb	34,399	35,899	37,656	39,620
(kg)	(15,603)	(16,284)	(17,080)	(17,971)
Combat weight, lb	51,345	60,690	58,289	62,089
(kg)	(23,290)	(27,529)	(26,439)	(28,163)
Max T.O. weight (carrier), lb	70,000	70,000	70,000	73,000
(kg)	(31,751)	(31,751)	(31,751)	(33,112)
Max. T.O. weight (land), lb	70,000	70,000	70,000	78,000
(kg)	(31,751)	(31,751)	(31,751)	(35,380)
Max. landing weight (carrier), lb	42,980	45,922	45,900	49,000
(kg)	(19,495)	(20,830)	(20,820)	(22,226)
Max. landing weight (land), lb	53,000	55,942	55,940	56,000
(kg)	(24,040)	(25,375)	(25,374)	(25,401)
Power loading @ combat wt., lb/lb st	2.65	3.03	3.07	2.96
(kg/kg st)	(2.65)	(3.03)	(3.07)	(2.96)
Max. power loading, lb/lb st	3.68	3.50	3.68	3.71
(kg/kg st)	(3.68)	(3.50)	(3.68)	(3.71)
Wing loading at combat wt., lb/sq ft	65.9	77.9	74.8	76.5
(kg/sq m)	(321.8)	(380.4)	(365.3)	(373.3)
Max. wing loading, lb/sq ft	89.9	89.9	89.9	96.1
(kg/sq m)	(438.7)	(438.7)	(438.7)	(469.0)
FUEL TANK CAPACITY:				
Fuselage fuel tanks, gallons	3,115	3,173	3,115	3,040
(liters)	(11,791)	(12,011)	(11,791)	(11,507)
Wing tanks, gallons	1,370	1,333	1,370	1,298
(liters)	(5,186)	(5,046)	(5,186)	(4,913)
Aux. bomb bay tanks, gallons	NA/NA	NA/NA	440/NA	748/NA
(liters)	(NA/NA)	(NA/NA)	(1,666/NA)	(2,831/NA)
Total capacity, gallons	4,485	4,506	4,925	5,086
(liters)	(16,977)	(17,057)	(18,643)	(19,251)

DIMENSIONS, WEIGHTS AND TANK CAPACITY

	A3D-2P	A3D-2Q	A3D-2T	KA-3B (CLE wing)
DIMENSIONS:				
Span, ft	72.5	72.5	72.5	72.5
(m)	(22.10)	(22.10)	(22.10)	(22.10)
Span, (wing folded), ft	49.2	49.2	49.2	49.2
(m)	(15.0)	(15.0)	(15.0)	(15.0)
Length/length with probe, ft	74.7/78.2	76.6/80.1	74.4/77.9	76.3/79.8
(m)	(22.77/23.84)	(23.35/24.41)	(22.68/23.74)	(23.26/24.32)
Height, ft	22.8	23.4	22.8	22.8
(m)	(6.95)	(7.13)	(6.95)	(6.95)
Height (tail folded), ft	15.9	16.1	15.2	15.5
(m)	(4.85)	(4.91)	(4.63)	(4.72)
Wing area, sq ft	779	779	812	812
(sq m)	(72.37)	(72.37)	(75.44)	(75.44)
WEIGHTS AND LOADINGS:				
Empty weight, lb	40,852	41,193	38,846	37,329
kg	(18,530)	(18,685)	(17,620)	(16,932)
Combat weight, lb	61,608	61,593	55,258	59,798
(kg)	(27,945)	(27,938)	(25,065)	(27,124)
Max T.O. weight (carrier), lb	73,000	73,000	70,000	70,000
(kg)	(33,112)	(33,112)	(31,751)	(31,751)
Max. T.O. weight (land), lb	78,000	78,000	84,000	84,000
(kg)	(35,380)	(35,380)	(38,012)	(38,012)
Max. landing weight (carrier), lb	49,000	49,000	47,057	45,922
(kg)	(22,226)	(22,226)	(21,345)	(20,830)
Max. landing weight (land), lb	56,000	56,000	55,942	55,942
(kg)	(25,401)	(25,401)	(25,375)	(25,375)
Power loading @ combat wt., lb/lb st	2.93	2.93	2.63	2.85
(kg/kg st)	(2.93)	(2.93)	(2.63)	(2.85)
Max. power loading, lb/lb st	3.71	3.71	4.0	4.0
(kg/kg st)	(3.71)	(3.71)	(4.0)	(4.0)
Wing loading @ combat wt., lb/sq ft	79.1	79.1	68.1	73.6
(kg/sq m)	(386.1)	(386.1)	(332.3)	(359.4)
Max. wing loading, lb/sq ft	100.1	100.1	103.4	103.4
(kg/sq m)	(488.9)	(488.9)	(505.1)	(505.1)
FUEL TANK CAPACITY:				
Fuselage fuel tanks, gallons	3,114	3,114	3,783	3,173
(liters)	(11,788)	(11,788)	(14,320)	(12,011)
Wing tanks, gallons	1,298	1,298	1,333	1,333
(liters)	(4,913)	(4,913)	(5,046)	(5,046)
Aux. bomb bay tanks, gallons	NA/NA	NA/NA	440/NA	2,109/NA
(liters)	(NA/NA)	(NA/NA)	(1,666/NA)	(7,983/NA)
Total capacity, gallons	4,412	4,412	5,116	6,615
(liters)	(16,701)	(16,701)	(19,366)	(25,040)

PERFORMANCE

	XA3D-1	A3D-1	YA3D-1P	A3D-2 (CLE wing)
SPEED:				
Max. speed at S.L., kn	564	534	532	556
(kmh)	(1,045)	(990)	(986)	(1,030)
Max. speed at altitude, kn at ft	---/---	509/35,000	524/18,000	508/35,000
(kmh at m)	(---/---)	(943/10,670)	(971/5,485)	(941/10,670)
Cruising speed, kn at ft	440/36,000	460/35,000	470/41,000	436/35,000
(kmh at m)	(815/10,975)	(852/10,670)	(871/12,495)	(808/10,670)
CLIMB RATE AND CEILING:				
Initial climb rate, ft/min	5,400/1	5,530/1	4,790/1	6,510/1
(m/min)	(1,646/1)	(1,686/1)	(1,460/1)	(1,984/1)
Time to climb, ft/min	30,000/8.9	30,000/10.1	30,000/11.3	30,000/8.4
(m/min)	(9,144/8.9)	(9,144/10.1)	(9,144/11.3)	(9,144/8.4)
Service ceiling, ft	41,000	39,000	41,800	40,400
(m)	(12,495)	(11,885)	(12,740)	(12,315)
PAYLOAD AND RANGE:				
Maximum weapons load, lb	8,700	12,800	NA	12,800
(kg)	(3,946)	(5,806)	(NA)	(5,806)
Bombload at combat radius, lb	8,700	7,600	NA	6,105
(kg)	(3,946)	(3,447)	(NA)	(2,769)
Combat radius, naut. miles	1,075	1,000	1,345	1,230
(km)	(1,991)	(1,852)	(2,493)	(2,280)
Unrefueled ferry range, naut. miles	----	2,611	----	----
(km)	(----)	(4,835)	(----)	(----)

PERFORMANCE

	A3D-2P	A3D-2Q	A3D-2T	KA-3B (CLE wing)
SPEED:				
Max. speed at S.L., kn	556	557	533	539
(kmh)	(1,030)	(1,032)	(987)	(999)
Max. speed at altitude, kn at ft	510/35,000	511/35,000	503/35,000	509/35,000
(kmh at m)	(945/10,670)	(947/10,670)	(932/10,670)	(943/10,670)
Cruising speed, kn at ft	459/35,000	459/35,000	430/35,000	430/35,000
(kmh at m)	(851/10,670)	(851/10,670)	(797/10,670)	(797/10,670)
CLIMB RATE AND CEILING:				
Initial climb rate, ft/min	6,150/1	6,150/1	4,810/1	5,620/1
(m/min)	(1,875/1)	(1,875/1)	(1,466/1)	(1,713/1)
Time to climb, ft/min	30,000/9.3	30,000/9.3	30,000/9.0	30,000/11.2
(m/min)	(9,144/9.3)	(9,144/9.3)	(9,144/9.0)	(9,144/11.2)
Service ceiling, ft	38,900	39,000	41,700	41,100
(m)	(11,855)	(11,885)	(12,710)	(12,525)
PAYLOAD AND RANGE:				
Maximum weapons load, lb	NA	NA	6,686	NA
(kg)	(NA)	(NA)	(3,033)	(NA)
Bombload at combat radius, lb	NA	NA	1,500	NA
(kg)	(NA)	(NA)	(680)	(NA)
Combat radius, naut. miles	1,110	1,110	1,122	900
(km)	(2,057)	(2,057)	(2,078)	(1,668)
Unrefueled ferry range, naut. miles	----	----	----	----
(km)	(----)	(----)	(----)	(----)

Chapter 10
Powerplant

The Pratt & Whitney J57-PW-10 turbojet engine has a maximum thrust rating of 10,500 lbs., a military thrust rating of 10,500 lbs., and a normal thrust rating of 9,000 lbs. It is a non-afterburning configuration noted primarily for its dependability and ruggedness. Well over 12,000 J57s were built by Pratt & Whitney before production was discontinued. Two views show both sides of a J57-PW-10 on standard transport dolly.

Engines—The KA-3B is powered by two Pratt & Whitney J57-P-10 turbojet engines installed in underwing nacelles. The J57-P-10 is a continuous flow, non-afterburning, gas turbine engine consisting of two multistage, axial flow compressors; eight combustion chambers; and a split, three-stage turbine assembly.

The axial flow dual compressor has a nine-stage low pressure unit and a seven-stage high pressure unit. The dual compressor supplies air to the can annular combustion chambers where fuel is instroduced and burned. The gas stream from the combustion section enters the turbine section giving energy to the split, three-stage turbines and driving the dual compressor. The first turbine stage drives the high pressure compressor by a hollow shaft. The second and third turbine stages drive the low pressure compressor by a concentric shaft through the hollow high pressure compressor (N2) shaft. A gearbox at the bottom of the engine provides external drive pads for the starter, tachometer, fuel pump, and fuel control units, and includes the engine oil pressure and scavenge pumps. The low pressure, or N1 rotor, has three tachometer drive pads at the bottom of the air inlet case. The engine has two automatically activated overboard bleed valves located between the low and high compressors, to facilitate starting, to improve acceleration, and to prevent surge by ducting low pressure air overboard during low power operation. High pressure compressor bleed air is used to close these bleed valves and to perform other engine service functions such as anti-icing and cooling of engine internal parts subject to extremely high temperatures. High pressure bleed air is also available for driving aircraft accessories and cabin pressurization.[2]

The combustion section contains eight burner cans. Fuel is sprayed into the burner cans through dual-orifice nozzles mounted in clusters of six at the inlet of each burner can. The fuel is ignited by two spark igniters located in burner cans No. 4 and 5, and the flame is propagated to the other burner cans by connecting flame tubes. The continuous combustion supplies the heat for expanding the compressor air, increasing the velocity of the gas flow to the turbine as described above. The turbine velocity gases are discharged through a fixed area exhaust nozzle.

Dimension:
Overall diameter........ 40.5 in. (1.03 m)
Length.............. 127.52 in. (3.24 m)
Weight:
4,200 lb. (1,905 kg)
Oil:
A 5.5-gallon (20.8-liter) oil tank is attached to each turbojet.

Performance ratings:

	Thrust	RPM	
		N1 compressor	N2 compressor
Maximum	10,500 lb. 4,763 kg	6,150	9,900
Military	10,500 lb. 4,763 kg	6,150	9,900
Normal	9,000 lb. 4,082 kg	5,900	9,650

A twelve-bottle JATO system, initially installed to provide the aircraft with additional thrust during airfield takeoff, has now been inactivated. Six MK 7 MOD 1 or MK 7 MOD 2 KS-4500 JATO bottles with igniters were mounted on each side of the fuselage aft of the bomb bay. Each bottle was capable of producing 4,500 lb. (2,041 kg) of thrust for five seconds. Due to the angle at which the bottles were installed, the maximum vectored thrust of the 12-bottle system was 45,000 lb. (20,412 kg). Combinations of two, four, or six units could be utilized depending on operational demands. The bottles were fired electrically and were jettisoned by means of utility hydraulic system pressure applied through a solenoid-operated valve.

Fuel System—As indicated earlier, the fuel system varies from one KA-3B to another, depending on whether the tanker package was fitted to the aircraft during factory production or added later during overhaul and modification at NARF Alameda. The following description applies to aircraft fitted with the tanker package during production by Douglas.

Three fuel tanks are located in the fuselage and one integral tank is installed in each wing inboard section. The fuselage tanks are self-sealing, but the wing tanks are of the non-self-sealing type. The fuel tanks are pressurized to a greater pressure than the atmosphere to prevent collapse of the tanks. This pressure is created by ram air from the fuel vent mast located on the left-hand side of the aft fuselage.

The forward fuselage tank is located ahead of the bomb bay and above the companionway; it has a design capacity of 1,225 U.S. gallons (4,637 liters) and a usable capacity of 1,170 gallons (4,429 liters). The auxiliary tank, in the upper bomb bay, has a design capacity of 793 gallons (3,002 liters) and a usable capacity of 748 gallons (2,831 liters). Immediately aft of the bomb bay, the third fuselage tank holds 1,928 gallons (7.298 liters) of which 1,870 gallons (7,079 liters) are usable. The wing tanks, located between the spars inboard of the wing fold, each have a design capacity of 668 gallons (2,529 liters) and a usable capacity of 649 gallons (2,547 liters). Total usable fuel capacity is thus 5,086 U.S. Gallons (19,252 liters). Some aircraft can be fitted to carry an unprotected tank containing 809 gallons (3,062 liters) in the bomb bay. JP-5 or JP-4 fuel is recommended for normal operations afloat or ashore. Low grade AVGAS is approved as an emergency fuel.

Fuel is supplied to the engines from the aft fuselage tank only. The forward fuselage tank and the auxiliary fuselage tank feed directly into the aft fuselage tank through a transfer system, which also functions as the fuel trim system. Fuel from the wing tanks must be transferred to the forward fuselage tank before being transferred to the aft fuselage tank, except in tanking operations.

The pressure fueling system is designed to permit fueling at approximately 250 gallons (946 liters) per minute through each of three pressure fueling receptacles, with separate receptacles and stop valves being provided for the forward fuselage tank, the wing tanks, and the aft tank. The aircraft is equipped with an air refueling probe on the left side of the forward fuselage. The probe is fitted with a Flight Refueling, Inc. MA-2 nozzle. The rate at which the receiver will accept fuel and the tanks into which fuel will be received is dependent upon receiver fuel remaining and fuel distribution. Transfer rate is dependent on tanker fuel state and distribution, with a maximum rate of 420 gallons (1,590 liters) per minute being attainable.

For its design mission, the KA-3B is fitted with a Flight Refueling, Inc. Model A-12B-7 Hose Reel Unit on the lower side of the fuselage at the aft end of the bomb bay. Approximately 61 ft. (18.59

---

[2] The extensive bleed air system has always been considered the weak point in the A-3. Nothing gets the attention of a crew faster than a big bleed air leak. An RVAH-3 TA-3B (BuNo 144861) crashed on August 27, 1973, largely due to a major leak in the cockpit.

m) of 2-1/8 in. (5.4 cm) diameter fueling hose is wound on the hydraulically operated reel inside the aft end of the bomb bay. A standard collapsible drogue (basket) is attached at the end of the hose; it provides the required aerodynamic drag for extension, stabilizes the hose, and acts as the target for the receiver's probe. Following completion of the refueling, and after the receiver has disengaged and cleared the tanker, the hose is rewound until the drogue is in the fairing beneath the fuselage of the tanker. If the refueling hose does not rewind, it is severed by means of an explosive cartridge hose cutter (guillotine) built in the Hose Reel Unit.

The altitude for air refueling may vary from sea level to more than 35,000 ft. (10,760 m); however, the desirable minimum altitude is 1,500 ft. (460 m). Refueling may be performed at airspeeds of from 220 KIAS (407 kmh) to 300 KIAS (556 kmh). *Skywarrior*-to-*Skywarrior* refueling is most comfortably accomplished at an altitude of 20,000 ft. (6,095 m) and an indicated airspeed of 250 KIAS (463 kmh). For refueling other types of aircraft, the altitude and airspeeds selected are those best suited to the receiver aircraft and within the operating limitations of the KA-3B.

During air refueling, fuel is transferred from the tanker's wing tanks into its auxiliary fuselage tank for transfer through the hose and drogue coupling into the refueling receptacle of the receiver aircraft. Transfer of fuel is controlled by the KA-3B navigator who has the tanker transfer control panel on his right side console. A two-color light signal system is incorporated in the tanker drogue fairing; an amber light indicates that the hose is in response condition and a green light, that the hose reel is in the fueling range and that fuel is flowing.

When all tanks of the tanker aircraft are installed and fully fueled, approximately 21,500 lb. (3,162 gallons/11,969 liters) of JP-5 fuel are available for refueling a receiver aircraft at a maximum fueling rate of 420 gallons (1,590 liters) per minute.

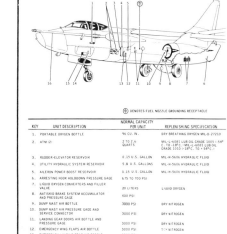

CAPACITIES

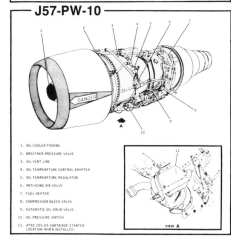

J57-PW-10

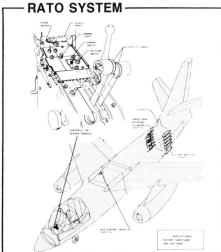

RATO SYSTEM

Ground maintenance of the J57-PW-10 is facilitated by its pylon mounting and easy accessibility. The engine is removable for more comprehensive overhaul work, being attached by a series of load-bearing lug mounts to the pylon mounts. The main oil tank is visible as a saddle over the engine compressor section.

The engine oil cooler is mounted in the intake bullet with airflow entering a centrally mounted intake and exhausting into the compressor face. Power take-off for auxiliary systems is transmitted through a shaft off a gearbox at the compressor face. The nacelle design is optimized to provide a short throat dimension.

The inflight refueling boom seen on most operational A-3s was added after the aircraft entered the fleet inventory. The original specification for the aircraft did not initially call for inflight refueling capability because it was assumed that, as a carrier borne bomber, the carrier would get it to well within target range. However, after the initial A3D configuration had been firmed up, the Navy adopted air refueling as a standard procedure to increase the range of most of its carrier aircraft.

The KA-3B is equipped with a hose and drogue-type inflight refueling system. The hose is wound onto a hydraulically-powered reel which is mounted at the aft end of the bomb bay. An interconnecting plumbing system interfaces the reel, pump assembly, and drogue with the bomb bay-mounted fuel tanks. Small lights, mounted at the trailing edge of the hose/drogue fairing pylon, provide visual clues during the refueling process.

The aft tank pressure refueling attachment point is located in the main gear well. Because of its location, it is easily reached by ground support personnel.

The emergency fuel dump pipe is located on the aft end of the fuselage, just below and forward of the rudder leading edge. The exit port of the dump pipe is designed to optimize both the dump flow rate and the dump plume. A small fence stabilizes span-wise airflow that otherwise might affect the plume pattern.

Appendix I: Individual Aircraft Histories

Revised February 20, 1986.

Seq. No.	FSN	BuNo	Model	Acc. Date	Disp. Date	Remarks & Disposition
1	7588	125412	XA3D-1	5-54	5-56	Diverted to ground training at Nav Tech Training Center, Memphis, TN.
2	7589	125413	XA3D-1			On display aboard the USS *Intrepid* in New York.
3	9253	130352	YA3D-1	10-53	4-66	Became NA-3A. Struck off at NARF Alameda following June '65 inflight accident while operated by NMC from Point Mugu.
4	9254	130353	A3D-1	11-53	10-64	At Edwards AFB in Sept. '54 for Navy Preliminary Performance and Fuel Consumption Tests. Tanker trials with A3D-2 (138903) during 1956. Lost in flight at El Centro in Oct. '64.
5	9255	130354	A3D-1	4-54	12-70	Salvaged at MASDC.
6	9256	130355	A3D-1	4-53	5-65	Flight tested in 1959 with J75-P-2 engines. Stored at Litchfield Park, AZ, in Aug. '64 and struck off there in May '65.
7	9257	130356	YA3D-1Q	7-54	10-58	Modified by Navy as YA3D-1Q. Lost near Port Lyautey, Morocco, while serving with VQ-2.
8	9258	130357	A3D-1	4-55	2-70	Served as NA-3A with NASWF. Service life expired in 1970 and salvaged at NARF Alameda.

130352 as an NA-3A at the Naval Missile Center. A cooling system is fitted in the tail with an external pipe carrying refrigerated air to the electronic equipment in the nose.

130357, with the Thunderbird tail markings of the Naval Weapons Evaluation Facility, photographed at NAS Alameda in November, 1969.

130363 as it appeared during the late sixties while operated as an NEA-3A by the Naval Missile Center at NAS Point Mugu.

135407 of VAH-123 landing at NAS Whidbey Island. This aircraft later went to the NATF Lakehurst in the markings of which it is shown on page 73.

135409 as an NA-3A of the Naval Missile Center in overall white finish enhanced by bright red nose, tail and wing tip markings.

135411 photographed at NAS Alameda in March, 1971, when it was fitted with a thimble nose radome while on bail to Hughes Aircraft Company.

135418 served at the Naval Missile Center until the end of 1974 prior to being donated to the Naval Aviation Museum at NAS Pensacola.

Seq. No.	FSN	BuNo	Model	Acc. Date	Disp. Date	Remarks & Disposition
9	9259	130358	YA3D-1P	6-54	12-64	Modified by Douglas as YA3D-1P and used for RDT&E at NADC Johnsville, PA. Became NRA-3A prior to being salvaged in Dec. '64.
10	9260	130359	A3D-1		10-55	Crashed near Edwards AFB on Oct. 29, '55.
11	9261	130360	A3D-1Q	3-55	1-58	Modified by Navy as A3D-1Q. Salvaged after inflight accident near Port Lyautey, Morocco, while serving with VQ-2.
12	9262	130361	A3D-1Q	3-55	12-70	Modified by Navy as A3D-1Q. Stored at MASDC in 1970. Donated to Pima County Air Museum.
13	9263	130362	A3D-1Q	7-55	5-59	Modified by Navy as A3D-1Q. Scrapped following inflight accident while serving with VQ-1 at Iwakuni, Japan.
14	9264	130363	A3D-1Q	8-55	1-70	Modified by Navy as A3D-1Q. Later used at Point Mugu as NEA-3A. Salvaged at Point Mugu R & D.
15	10300	135407	A3D-1	1-55	7-71	Salvaged at MASDC.
16	10301	135408	A3D-1	7-55	1-60	At NATC, as YA3D-1, for carrier suitability trials from Sept. '56 until June '57. Salvaged after inflight accident while serving with VAH-123.
17	10302	135409	A3D-1	7-55	5-75	Operated at one time as NA-3A. Salvaged at MASDC.
18	10303	135410	A3D-1	5-55	1-59	Missing on a flight from Point Mugu.
19	10304	135411	A3D-1	8-55	Stored	Bailed to Hughes as testbed. In storage at MASDC since Jan. '75.
20	10305	135412	A3D-1	8-55	8-70	Diverted to ground training at Lakehurst, NJ.
21	10306	135413	A3D-1	10-55	9-63	Initially assigned to NASWF for special weapons tests. Lost in flight accident from NARF Alameda.
22	10307	135414	A3D-1	12-55	11-65	Lost in flight accident from Lakehurst, NJ.
23	10308	135415	A3D-1	1-56	4-69	Salvaged at NARF Alameda.
24	10309	135416	A3D-1	5-56	10-63	Salvaged following inflight accident while serving with VAH-3.
25	10310	135417	A3D-1	5-56	9-57	Lost in flight while serving with VAH-1.
26	10311	135418	A3D-1	7-56	12-74	Damaged and later donated to the Naval Aviation Museum at NAS Pensacola.
27	10312	135419	A3D-1	7-56	3-61	Salvaged following inflight accident while serving with VAH-123.
28	10313	135420	A3D-1	6-56	10-64	Stored at Litchfield Park, AZ. Final disposition unknown.
29	10314	135421	A3D-1	1-56	8-64	Diverted to ground training at Lakehurst, NJ.
30	10315	135422	A3D-1	1-56	5-60	Lost in flight accident while serving with VAH-3.
31	10316	135423	A3D-1	1-56	5-69	Diverted to ground training at NARF Alameda.
32	10317	135424	A3D-1	2-56	2-56	Salvaged at Patuxent River in Feb. '56 following crash due to loss of main landing gear door during delivery flight.
33	10318	135425	A3D-1	5-56	12-63	At NATC from March '56 until June '57 for fuel consumption tests. Lost in flight accident from NARF Norfolk.
34	10319	135426	A3D-1	5-56	11-63	Lost in flight accident while operated by VAH-123.
35	10320	135427	A3D-1	6-56	8-70	Became NA-3A for use in *Phoenix* missile tests. Salvaged at NARF Alameda.

Seq. No.	FSN	BuNo	Model	Acc. Date	Disp. Date	Remarks & Disposition
36	10321	135428	A3D-1	7-56	12-67	Salvaged at NARF Alameda.
37	10322	135429	A3D-1	7-56	6-59	Lost in flight accident while operated by VAH-1.
38	10323	135430	A3D-1	10-56	6-57	Salvaged at Jacksonville after flight accident while operated by VAH-3.
39	10324	135431	A3D-1	8-56	1-68	Salvaged at NARF Alameda.
40	10325	135432	A3D-1	9-56	8-68	Salvaged at NARF Alameda.
41	10326	135433	A3D-1	10-56	3-64	Diverted to ground training at NWL Dahlgreen, VA. Later went to NAS Norfolk, VA, and NAS Key West, FL, for use by VAQ-33 FRAMP.
42	10327	135434	A3D-1	11-55	6-67	Damaged on the ground and initially stored at MASDC. Later shipped to Edwards AFB for use in fire-fighting training.
43	10328	135435	A3D-1	2-56	4-66	Diverted to ground training at Lakehurst, NJ.
44	10329	135436	A3D-1	2-56	1-68	Salvaged at NARF Alameda.
45	10330	135437	A3D-1	8-56	2-57	First operational loss. Both engines flamed out while aircraft was approaching Mayport, FL, on Feb. 9, '57. VAH-3 crew bailed out but pilot's parachute did not open. Aircraft came down intact in shallow water and was subsequently salvaged.
46	10331	135438	A3D-1	5-56	8-68	Salvaged at NARF Alameda.
47	10332	135439	A3D-1	6-56	10-64	Stored at Litchfield Park, AZ. Final disposition unknown.
48	10333	135440	A3D-1	4-56	8-59	Lost in flight accident while operated by VAH-1.
49	10334	135441	A3D-1	8-56	8-57	Lost in flight accident while operated by VAH-3.
50	10335	135442	A3D-1	9-56	3-63	Lost in flight accident while operated by VAH-123.
51	10336	135443	A3D-1	10-56	3-64	Diverted to ground training at NWL Dahlgreen, VA.
52	10337	135444	A3D-1	8-56	6-67	Salvaged at MASDC.
53	10763	138902	A3D-2	5-56	8-60	Lost during flight from USS *Independence* with VAH-1.
54	10764	138903	A3D-2	4-56	2-58	Tanker trials with A3D-1 (130353) during 1956. Crashed at Edwards AFB in Jan. '58 during autopilot tests.
55	10765	138904	A3D-2	12-56	5-75	At NATC from Nov. '57 until June '59 for fuel consumption tests, then carrier suitability trials (Dec. '57, and March-April '58) and carquals aboard USS *Forrestal* in Apr. '58. Converted to KA-3B by NARF Alameda in Sept. '57. Salvaged at MASDC.
56	10766	138905	A3D-2	10-56	5-75	At NATC, as YA3D-2, for combined Stability and Control, and Aircraft and Engine Performance Trials from Nov. '57 until Apr. '58. Converted to KA-3B by NARF Alameda in Sept. '67. Still in SARDIP.
57	10767	138906	A3D-2	3-57	5-75	Converted to KA-3B by NARF Alameda in June '57. Still in SARDIP.
58	10768	138907	A3D-2	3-57	8-63	Lost during flight from USS *Constellation* with VAH-10.
59	10769	138908	A3D-2	11-56	1-58	Crashed on launch from USS *Ticonderoga* during operations in the South China Sea with VAH-2.
60	10770	138909	A3D-2	11-56	12-68	Converted to KA-3B by NARF Alameda in July '57. Crashed on launch from USS *Enterprise* on Dec. 2, '68 while operated by VAQ-132.
61	10771	138910	A3D-2	11-56	8-57	Lost during flight from NAS Whidbey Island with VAH-4.

135427, the primary "Phoenix Missile System" testbed, fitted with the F-14's Hughes AWG-9 fire control system. NAS Alameda, June, 1968.

138902, the first A3D-2, on display at Edwards AFB on the occasion of Armed Forces Day in May, 1957. The protrusion in the nose is an instrumentation boom.

138908, one of the first A3D-2s assigned to VAH-2, deployed with Det Mike aboard the USS "Ticonderoga" (CVA-14) in 1957-58.

138918, an A-3B of VAH-11 landing at NAS Sanford, Florida. Assigned to CVW-1, it had just returned from a deployment aboard the USS "Franklin D. Roosevelt (CVA-42).

138925, an A3D-2 from VAH-9, aboard the USS "Saratoga" (CVA-60). It is still being used as a FRAMP training airplane at NAS Key West.

138928 during early tanker trials in 1957. Converted as KA-3B ten years later and salvaged in 1970.

138931, A KA-3B of VAQ-131, at NAS Alameda on October 23, 1968, prior to a deployment aboard the USS "Kitty Hawk" (CVA-63).

138933 at MASDC on March 19, 1974. Still kept in SARDIP at the Aerospace Maintenance and Regeneration Center (AMARC), formerly MASDC.

138939, an A3D-2 of VAH-9 aboard the USS "Saratoga" (CVA-60) in 1958. Currently with PMTC at NAS Point Mugu as an NA-3B.

138943, an A3D-2 from VAH-11, bolting off the USS "Franklin D. Roosevelt" (CVA-42) on June 21, 1961, during operations in the Med.

Seq. No.	FSN	BuNo	Model	Acc. Date	Disp. Date	Remarks & Disposition
62	10772	138911	A3D-2	2-57	5-75	Converted to KA-3B by NARF Alameda in May '69. Still in SARDIP.
63	10773	138912	A3D-2	12-56	8-60	Salvaged after accident aboard USS *Oriskany* while operated by VAH-4.
64	10774	138913	A3D-2	12-56	5-59	Lost during flight from NAS Whidbey Island with VAH-4.
65	10775	138914	A3D-2	11-56	5-75	Converted to KA-3B by NARF Alameda in June '67. Still in SARDIP.
66	10776	138915	A3D-2	12-56	5-75	Converted to KA-3B by NARF Alameda in Sept. '68. Scrapped at MASDC.
67	10777	138916	A3D-2	1-57	7-63	Salvaged following inflight accident while operated from NAS Whidbey Island by VAH-6.
68	10778	138917	A3D-2	12-56	4-67	Inflight accident while operated from NAS Whidbey Island by VAH-123. Final disposition unknown.
69	10779	138918	A3D-2	6-57	1-69	Prototype CLE installation. Converted by NARF Alameda to KA-3B in June '67 and to EKA-3B in Jan. '68. Destroyed in Jan. 14, '69 fire aboard USS *Enterprise*.
70	10780	138919	A3D-2	12-56	5-57	Crashed during carquals aboard USS *Bon Homme Richard* while operated by VAH-2.
71	10781	138920	A3D-2	12-56	6-67	Service life expired. Sent to MASDC. Final disposition unknown, presumed salvaged.
72	10782	138921	A3D-2	1-57	4-68	Converted to KA-3B by NARF Alameda in June '67. Lost in flight accident while operated from NAS Whidbey Island by VAH-2.
73	10783	138922	A3D-2	2-57	10-73	Converted to special electronic configuration in July '67 and operated for the Navy by McDonnell Douglas; later assigned to FEWSG. Diverted to ground training at Lakehurst, NJ. Struck off in Oct. '73.
74	10784	138923	A3D-2	2-57	6-73	Converted to KA-3B by NARF Alameda in Jan. '69. Salvaged after normal wear and tear.
75	10785	138924	A3D-2	1-57	7-61	Lost in flight accident while operating from USS *Bon Homme Richard* with VAH-4.
76	10786	138925	A3D-2	3-57	Active	Converted to KA-3B by NARF Alameda in Apr. '69. Currently as FRAMP with VAQ-33.
77	10787	138926	A3D-2	3-57	3-59	Salvaged after accident aboard USS *Forrestal* while operated by VAH-5.
78	10788	138927	A3D-2	4-57	2-63	Lost in flight accident while operated by VAH-4 from USS *Ticonderoga*.
79	10789	138928	A3D-2	2-57	7-70	Converted to KA-3B by NARF Alameda in July '67. Time expired and salvaged.
80	10790	138929	A3D-2	3-57	1-79	Converted to KA-3B by NARF Alameda in Sept. '68. Crashed at Patuxent River on Jan. 26, '79.
81	10791	138930	A3D-2	4-57	2-65	Lost in flight accident on Feb. 12, '65 while operated from NAS Whidbey Island by VAH-4.
82	10792	138931	A3D-2	5-57	5-75	Converted to KA-3B by NARF Alameda in May '68. Still in SARDIP.
83	10793	138932	A3D-2	2-57	Stored	Prototype CLE installation. Stored at MASDC.
84	10794	138933	A3D-2	4-57	5-75	Converted to KA-3B by NARF Alameda in Jan '69. Still in SARDIP.

Seq. No.	FSN	BuNo	Model	Acc. Date	Disp. Date	Remarks & Disposition
85	10795	138934	A3D-2	5-57	3-63	Lost in flight accident while operated from NAS Whidbey Island by VAH-13.
86	10796	138935	A3D-2	5-57	8-58	Damaged in accident while operated by VAH-4 from USS *Shangri-La* in Aug. '58. Subsequently sent to MASDC and eventually salvaged.
87	10797	138936	A3D-2	5-57	10-57	Lost off Cubi Point while operated by VAH-2.
88	10798	138937	A3D-2	5-57	7-69	Converted to KA-3B by NARF Alameda in Jan. '68. Struck off following inflight accident while operated from NAS Whidbey Island by VAH-123.
89	10799	138938	A3D-2	9-57	Active	With PMTC at Point Mugu as NA-3B.
90	10800	138939	A3D-2	6-57	7-76	Converted to KA-3B by NARF Alameda in June '67. Still in SARDIP.
91	10801	138940	A3D-2	6-57	2-70	Converted to KA-3B by NARF Alameda in June '67. Salvaged following inflight accident while operated from NAS Whidbey Island by VAH-123.
92	10802	138941	A3D-2	6-57	1-71	Converted to KA-3B by NARF Alameda in June '68. Time expired and salvaged at NARF Norfolk.
93	10803	138942	A3D-2	6-57	2-58	Damaged in accident while operated from NAS Sanford by VAH-9 in Feb. '58. Subsequently sent to MASDC and eventually salvaged.
94	10804	138943	A3D-2	7-57	2-69	Converted to KA-3B by NARF Alameda in March '68. Missing on Feb. 18, '69 (probably due to pilot's fatigue) while operated by VAH-2 from USS *Coral Sea* over Gulf of Tonkin during tanker sortie.
95	10805	138944	A3D-2	7-57	Active	Converted to KA-3B by NARF Alameda in Sept. '67. Currently with VAQ-34.
96	10806	138945	A3D-2	7-57	1-72	Converted to KA-3B by NARF Alameda in March '68. Salvaged at MASDC.
97	10807	138946	A3D-2	7-57	1-72	Converted to KA-3B by NARF Alameda in June '67. Salvaged at MASDC.
98	10808	138947	A3D-2	12-56	5-65	Lost on May 25, '65 due to structural failure on catapult launch while operated on tanker sortie by VAH-4 from the USS *Oriskany* in the Gulf of Tonkin.
99	10809	138948	A3D-2	7-57	8-60	Lost in flight accident while operated by VAH-9 from USS *Saratoga*.
100	10810	138949	A3D-2	8-57	8-58	Struck off after inflight accident while operated from NAS Sanford by VAH-11.
101	10811	138950	A3D-2	8-57	3-59	Struck off after accident aboard USS *Forrestal* while operated by VAH-5.
102	10812	138951	A3D-2	8-57	8-70	Converted to KA-3B by NARF Alameda in Sept. '67. Diverted to ground training.
103	10813	138952	A3D-2	9-57	3-67	Salvaged after accident while used for RDT&E at China Lake.
104	10814	138953	A3D-2	9-57	5-75	Converted to KA-3B by NARF Alameda in July '67. Salvaged at MASDC.
105	10815	138954	A3D-2	4-57	2-58	Lost during flight from USS *Saratoga* while operated by VAH-9.
106	10816	138955	A3D-2	4-57	4-70	Converted to KA-3B by NARF Alameda in June '67. Salvaged at NARF Alameda after inflight accident.

138944 was stored at MASDC for three years beginning in February, 1976, after serving as a KA-3B with NMC. Now operated by VAQ-34 as illustrated on page 65.

138953, an A3D-2 from VAH-1, about to be catapulted off the angle deck of the USS "Independence" (CVA-62).

138968, an NA-3B used as an electronic aggressor by VAQ-33, as it appeared shortly before being lost on October 10, 1972.

138974, an A-3B from VAH-11 Det 8, landing aboard the USS "Independence" (CVA-62) on October 20, 1965.

142238, a KA-3B from VAQ-308, at NAF Andrews on March 27, 1971, ten months after this reserve squadron had been commissioned.

142246, a KA-3B last used by the Weapons Service Test Division at NATC before being donated to the Bradley Aviation Museum.

142251, an A-3B of VAH-4, became a KA-3B in August, 1967, and an EKA-3B in July, 1968, and went to MASDC in April, 1982.

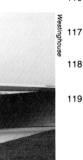

142256 first became an NRA-3B in 1963. For the past several years, it has been bailed to Westinghouse for use as a testbed.

142404, an EKA-3B from VAQ-132, at NAS Alameda in November, 1969. It was last operated by VAQ-135 before being sent to MASDC.

142630, an A3D-2 from VAH-8, at NAS Alameda on October 29, 1960. Now designated NA-3B, it is assigned to the NWC China Lake and illustrated on page 73.

Seq. No.	FSN	BuNo	Model	Acc. Date	Disp. Date	Remarks & Disposition
107	10817	138956	A3D-2	4-57	7-57	Scrapped after inflight accident while operated from USS *Saratoga* by VAH-9.
108	10818	138957	A3D-2	4-57	5-73	Converted to KA-3B by NARF Alameda in Sept. '67. Scrapped after inflight accident while operating from NARF Alameda.
109	10819	138958	A3D-2	5-57	4-60	Lost during flight from USS *Ranger* while operated by VAH-6.
110	10820	138959	A3D-2	5-57	1-72	Converted to KA-3B by NARF Alameda in Sept. '67. Salvaged at MASDC.
111	10821	138960	A3D-2	5-57	1-61	Lost in flight accident while operated from NAS Sanford by VAH-13.
112	10822	138961	A3D-2	6-57	1-72	Converted to KA-3B by NARF Alameda in June '67. Salvaged at MASDC.
113	10823	138962	A3D-2	6-57	6-62	Lost in flight accident while operated from NAS Sanford by VAH-11.
114	10824	138963	A3D-2	9-56	8-70	Converted to KA-3B by NARF Alameda in Dec. '67. Time expired and diverted to ground training at NARF Alameda.
115	10825	138964	A3D-2	6-57	Stored	Converted to KA-3B by NARF Alameda in Oct. '68. Stored at MASDC.
116	10826	138965	A3D-2	7-57	Stored	Converted to KA-3B by NARF Alameda in June '68. Stored at MASDC.
117	10827	138966	A3D-2	7-57	2-74	Converted to KA-3B by NARF Alameda in Aug. '68. Salvaged at NARF Alameda.
118	10828	138967	A3D-2	8-57	5-75	Converted to KA-3B by NARF Alameda in May '67. Still in SARDIP.
119	10829	138968	A3D-2	8-57	10-72	Converted to special electronic configuration in July, 1967, and operated for the Navy by McDonnell Douglas prior to transfer to FEWSG. Lost in flight accident on Oct. 10, '72 while operated by VAQ-33 from NAS Norfolk.
120	10830	138969	A3D-2	8-57	5-75	Converted to KA-3B by NARF Alameda in Jan. '69. Still in SARDIP.
121	10831	138970	A3D-2	9-57	6-60	Salvaged after inflight accident while operated from NAS Whidbey Island by VAH-4.
122	10832	138971	A3D-2	9-57	1-72	Converted to KA-3B by NARF Alameda in June '67. Salvaged at MASDC.
123	10833	138972	A3D-2	9-57	10-68	Converted to KA-3B by NARF Alameda in June '67. Lost in flight accident while operated from NAS Whidbey Island by VAH-10.
124	10834	138973	A3D-2	9-57	5-75	Converted to KA-3B by NARF Alameda in June '67. Still in SARDIP.
125	10835	138974	A3D-2	8-57	7-70	Converted to KA-3B by NARF Alameda in June '67. Salvaged at NARF Alameda.
126	10836	138975	A3D-2	8-57	4-68	Salvaged, NPRO Long Beach.
127	10837	138976	A3D-2	9-57	3-61	Lost in flight while operated from USS *F. D. Roosevelt* by VAH-11.
128	11562	142236	A3D-2	10-57	11-62	Lost in flight while operated from USS *Bon Homme Richard* by VAH-4.
129	11563	142237	A3D-2	10-57	5-75	Converted by NARF Alameda to KA-3B in June '67 and to EKA-3B in May '69. Scrapped at MASDC.

Seq. No.	FSN	BuNo	Model	Acc. Date	Disp. Date	Remarks & Disposition
130	11564	142238	A3D-2	10-57	5-78	Converted to KA-3B by NARF Alameda in June '67. Still in SARDIP.
131	11565	142239	A3D-2	10-57	1-73	Converted by NARF Alameda to KA-3B in June '67 and to EKA-3B in Jan. '69. Found beyond economical repair while undergoing rework and salvaged.
132	11566	142240	A3D-2	11-57	4-63	Damaged in flight and stored. Now on display aboard USS *Yorktown* in Charleston, SC.
133	11567	142241	A3D-2	10-57	10-70	Converted to KA-3B by NARF Alameda in Jan. '68. Salvaged at NARF Alameda.
134	11568	142242	A3D-2	12-57	4-76	Destroyed at Lakehurst, NJ, when forward and aft fuel cells exploded during off center arresting gear test.
135	11569	142243	A3D-2	11-57	1-62	Lost in flight accident while operated from NAS Sanford by VAH-1.
136	11570	142244	A3D-2	12-57	8-71	Converted to KA-3B by NARF Alameda in June '67. Service life expired while serving with VAQ-130 at NAS Alameda and salvaged.
137	11571	142245	A3D-2	12-57	5-61	Lost in flight while operated from USS *F. D. Roosevelt* by VAH-11.
138	11572	142246	A3D-2	11-57	2-76	Last served at NATC Patuxent River. Donated to Bradley Air Museum, CT.
139	11573	142247	A3D-2	12-57	5-78	Converted to KA-3B by NARF Alameda in June '67. Still in SARDIP.
140	11574	142248	A3D-2	1-58	3-75	Converted by NARF Alameda to KA-3B in June '67, to EKA-3B in Oct. '68 and back to KA-3B in March '69. Salvaged after inflight accident while used for flight training at NARF Alameda.
141	11575	142249	A3D-2	1-58	Stored	Converted by NARF Alameda to KA-3B in May '67 and to EKA-3B in July '68. Stored at MASDC since April '82.
142	11576	142250	A3D-2	1-58	12-64	Lost during training flight (due to control system failure) on Dec. 27, '64 while operated by VAH-4 from USS *Hancock* in Gulf of Tonkin.
143	11577	142251	A3D-2	2-58	Stored	Converted by NARF Alameda to KA-3B in Aug. '67 and to EKA-3B in June '69. Stored at MASDC since April '82.
144	11578	142252	A3D-2	2-58	8-70	Converted by NARF Alameda to KA-3B in July '67 and to EKA-3B in June '69. Crashed in Colorado following loss of control at high altitude while operated from NAS Alameda by VAQ-131.
145	11579	142253	A3D-2	1-58	5-78	Converted to KA-3B by NARF Alameda in June '67. Still in SARDIP.
146	11580	142254	A3D-2	3-58	3-59	Lost in flight while operated by VAH-6 from USS *Ranger*.
147	11581	142255	A3D-2	2-58	Stored	Converted by NARF Alameda to KA-3B in July '67 and to EKA-3B in Sept. '68. Stored at MASDC since May '75.
148	11582	142256	YA3D-2P	4-58	Active	Bailed to Douglas for combined Stability and Control, and Aircraft and Performance Trials at NATC in May/June '59. Modified as NRA-3B by Douglas in 1963 to provide operational control and airborne support for Project PRESS. Later fitted with experimental sonobuoys installation on fuselage side. Subsequently, bailed to Westinghouse. Still current in this role.

142632, an EKA-3B from VAQ-131, at NAS Alameda on April 15, 1970. As part of CVW-1, VAQ-132 deployed aboard the USS "John F. Kennedy" (CVA-67) in 1970-71.

142638, a KA-3B from VAH-10 Det 59, deployed to the Med aboard the USS "Forrestal" (CVA-59) from December 2, 1969, until July 8, 1970.

142646, an EKA-3B of VAQ-135 Det 3, was sent to MASDC shortly after returning from a cruise aboard the USS "Coral Sea" (CVA-43) on November 8, 1973.

142649, converted to KA-3B in July, 1967, photographed on March 27, 1971 at NAF Andrews.

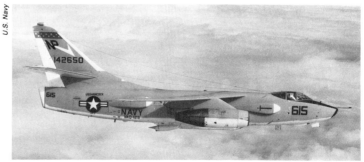

142650, an EKA-3B of VAQ-129 from the USS "Hancock" (CVA-19), in flight over the Gulf of Tonkin on March 18, 1971.

142651, a tanker-configured A3D-2 of VAH-9, refueling an F4H-1F of VF-101 Det A during "Operationa LANA" on May 12, 1961.
U.S. Navy

142652, stored at MASDC since August 23, 1973, last served with VAQ-135 Det 2 aboard the USS "Forrestal" (CVA-59).
Peter B. Lewis

142655 was lost in a launch accident aboard the USS "Oriskany" (CVA-34) on October 21, 1967. Photographed at Elmendorf AFB on April 22, 1964.
Norm Taylor

142656 at NAS Sanford on October 4, 1965. It had last deployed aboard the USS "Franklin D. Roosevelt" (CVA-42) between April 28 and December 22, 1964.
Ken Buchanan collection

142664, seen here in VAK-308 markings, now serves with the co-located VAK-208 to provide tanking and pathfinding for CVWR-20.
Aerofax, Inc. collection

Seq. No.	FSN	BuNo	Model	Acc. Date	Disp. Date	Remarks & Disposition
149	11583	142257	YA3D-2Q	8-58	3-74	Lost while operated by VQ-2.
150	11584	142400	A3D-2	3-58	7-70	Converted by NARF Alameda as KA-3B in May '67 and to EKA-3B in June '69. Crashed on July 4, 1970 after drag chute deployed during bolter when landing aboard the USS *America* in the Gulf of Tonkin at the end of a tanker sortie with VAQ-132.
151	11585	142401	A3D-2	2-58	5-78	Converted to KA-3B by NARF Alameda in June '67. Still in SARDIP.
152	11586	142402	A3D-2	3-58	9-62	Lost on Sept. 24, '62 in landing accident aboard USS *Midway* while operated by VAH-8.
153	11587	142403	A3D-2	4-58	Stored	Converted by NARF Alameda to KA-3B in June '67 and to EKA-3B in April '68. To NAVPRO Dallas for wing and catapult tests in Aug. '85.
154	11588	142404	A3D-2	3-58	Stored	At NATC during 1966 for Approach Power Compensator evaluation. Converted by NARF Alameda to KA-3B in July '69 and to EKA-3B in ??. Stored at MASDC since Dec. '73.
155	11589	142405	A3D-2	4-58	8-64	Damaged in flight while operated from NAS Whidbey Island by VAH-8 and salvaged.
156	11590	142406	A3D-2	4-58	1-72	Converted to KA-3B by NARF Alameda in June '67. Salvaged at MASDC.
157	11591	142407	A3D-2	4-58	3-61	Damaged beyond economical repair while operated aboard USS *Coral Sea* by VAH-2.
158	11693	142630	A3D-2	5-58	Active	Assigned to NPTR El Centro and then to NWC China Lake as NA-3B.
159	11694	142631	A3D-2	5-58	7-64	Lost in flight while operated from USS *F. D. Roosevelt* by VAH-11.
160	11695	142632	A3D-2	5-58	Stored	Converted by NARF Alameda to KA-3B in July '67 and to EKA-3B in June '68. Stored at MASDC since Apr. '82.
161	11696	142633	A3D-2	6-58	10-66	Lost on Oct. 2, '66 after shedding bridle when catapulted off USS *Coral Sea* in Gulf of Tonkin at start of ferry flight by VAH-2.
162	11697	142634	A3D-2	5-58	1-73	Converted to KA-3B by NARF Alameda in June '67 and later to EKA-3B. Lost on Jan. 21, '73 when catapulted from the USS *Ranger* in the Gulf of Tonkin while operated by VAQ-130 on tanker sortie.
163	11698	142635	A3D-2	6-58	1-72	Converted to KA-3B by NARF Alameda in June '67. Salvaged at MASDC.
164	11699	142636	A3D-2	6-58	12-58	Lost over Caribbean while operated by VAH-11.
165	11700	142637	A3D-2	7-58	10-61	Lost in flight while operated from NAS Sanford by VAH-11.
166	11701	142638	A3D-2	7-58	Stored	Converted by NARF Alameda to KA-3B in June '67 and to EKA-3B in Jan. '69. Stored at MASDC since April '82.
167	11702	142639	A3D-2	8-58	7-59	Lost in flight while operated from USS *Ranger* by VAH-6.
168	11703	142640	A3D-2	6-58	10-59	Crashed near Napa, CA, on Sept. 18, '59 while operated from NAS Whidbey Island by VAH-2.
169	11704	142641	A3D-2	7-58	7-59	Lost over Caribbean while operated by VAH-11.
170	11705	142642	A3D-2	7-58	3-60	Crashed 10 miles south of Harrington, WA, while operated from NAS Whidbey Island by VAH-2.

Seq. No.	FSN	BuNo	Model	Acc. Date	Disp. Date	Remarks & Disposition
171	11706	142643	A3D-2	8-58	4-59	Lost in flight while operated from NAS Sanford by VAH-7.
172	11707	142644	A3D-2	9-58	5-78	Converted to KA-3B by NARF Alameda in June '67. Still in SARDIP.
173	11708	142645	A3D-2	8-58	10-58	Lost in flight while operated from NAS Jacksonville by VAH-1.
174	11709	142646	A3D-2	9-58	Stored	Converted to EKA-3B by NARF Alameda in Sept. '68. Stored at MASDC since April '82.
175	11710	142647	A3D-2	9-58	Stored	Converted by NARF Alameda to KA-3B in May '67 and to EKA-3B in Nov. '68. Stored at MASDC since Nov. '73.
176	11711	142648	A3D-2	10-58	10-61	Lost in flight while operated from NAS Sanford by VAH-11.
177	11712	142649	A3D-2	12-58	10-72	Converted to KA-3B by NARF Alameda in July '67. Crashed at Buckley ANGB, CO, while operated from NAS Alameda by VAQ-308.
178	11713	142650	A3D-2	12-58	Active	Beginning with this aircraft all A3D-2s were delivered by Douglas with the tanker package. Converted by NARF Alameda to KA-3B in June '67 and to EKA-3B in June '68. Stored at MASDC in Apr. '82. Returned to service with VAQ-33 as KA-3B in Apr. '85.
179	11714	142651	A3D-2	3-59	Stored	Converted by NARF Alameda to KA-3B in June '67 and to EKA-3B in Nov. '67. Stored at MASDC since March '78.
180	11715	142652	A3D-2	3-59	Stored	Converted by NARF Alameda to KA-3B in June '67 and to EKA-3B in Nov. '67. Stored at MASDC since Aug. '73.
181	11716	142653	A3D-2	3-59	4-66	Shot down by MiG over China Sea on Apr. 12, '66 while ferried from Cubi Point to USS *Kitty Hawk* by VAH-4 crew.
182	11717	142654	A3D-2	4-59	Stored	Converted by NARF Alameda to KA-3B in June '67 and to EKA-3B in Nov. '67. Stored at MASDC since May '75.
183	11718	142655	A3D-2	3-59	10-67	Lost on Oct. 21, '67 due to accidental ignition of JATO bottles during launch from USS *Oriskany* in the Gulf of Tonkin while operated by VAH-4 on logistics flight.
184	11719	142656	A3D-2	3-59	Stored	Converted by NARF Alameda to KA-3B in June '67 and to EKA-3B in Jan. '68. Stored at MASDC since May '75.
185	11720	142657	A3D-2	3-59	5-70	Converted by NARF Alameda to KA-3B in July '67 and to EKA-3B in Dec. '67. Assigned to VAQ-135, this EKA-3B bingoed to Da Nang but was lost before reaching its destination during flight from Cubi Point to USS *Coral Sea* on May 16, '70.
186	11721	142658	A3D-2	3-59	7-67	Converted to KA-3B by NARF Alameda in June '67. Lost on July 28, '67 following double engine failure during tanker sortie while operated by VAH-4 from USS *Oriskany* in the Gulf of Tonkin.
187	11722	142659	A3D-2	4-59	7-77	Converted by NARF Alameda to KA-3B in June '67 and to EKA-3B in Sept. '67. Destroyed on July 10, '77 due to mishandled aborted take-off while operated by VAQ-308 from NAS Alameda.

142665, an A3D-2 of VAH-1, trapping aboard the USS "Saratoga" (CVA-60). It was lost off the USS "Enterprise" (CVAN-65) on April 1, 1966, while serving with VAH-4 Det M.

142667, an NRA-3B still operated in early 1987 by the Pacific Missile Test Center at NAS Point Mugu.

144628, a tanker-configured A-3B of VAH-4 Det B, deployed to the Gulf of Tonkin aboard the USS "Hancock" (CVA-19) from January 5 to July 22, 1967.

144825, a unique NRA-3B characterized by the large radome fitted by Grumman in 1960, has been operated by NMC/PMTC since 1963.

144827, an ERA-3B still operated by VAQ-33 in early 1987, starting its take-off roll at NAS Miramar on October 4, 1977.

Seq. No.	FSN	BuNo	Model	Acc. Date	Disp. Date	Remarks & Disposition
188	11723	142660	A3D-2	4-59	Stored	Converted by NARF Alameda to KA-3B in June '67 and to EKA-3B in Feb. '68. Stored at MASDC since May '78.
189	11724	142661	A3D-2	4-59	Stored	Converted by NARF Alameda to KA-3B in June '67 and to EKA-3B in Dec. '67. Stored at MASDC since May '78.
190	11725	142662	A3D-2	4-59	Active	Converted by NARF Alameda to KA-3B in ??, to EKA-3B in ?? and back to KA-3B in ??. Assigned to VAK-308.
191	11726	142663	A3D-2	5-59	10-61	Lost in flight while operated from NAS Sanford by VAH-5.
192	11727	142664	A3D-2	5-59	Active	Converted by NARF Alameda to KA-3B in July '67, to EKA-3B in Feb. '68 and back to KA-3B in Oct. '74. Assigned to VAK-208.
193	11728	142665	A3D-2	5-59	4-66	Lost on Apr. 1, '66 after nosewheel collapsed when catapulted from USS *Enterprise* in Gulf of Tonkin while operated by VAH-4 on tanker sortie.
194	11729	142666	A3D-2P	8-58	Active	Stored at MASDC but being reactivated in 1985.
195	11730	142667	A3D-2P	9-58	Active	Converted to NRA-3B and assigned to PMTC.
196	11731	142668	A3D-2P	10-58	Active	Stored at MASDC for several years and retrieved from storage in 1981 for conversion by NARF Alameda and assignment to VAQ-34 as ERA-3B.
197	11732	142669	A3D-2P	11-58	Active	Stored at MASDC but being reactivated in 1985.
198	11733	142670	A3D-2Q	9-58	6-68	Lost in flight accident while operated by VQ-2.
199	11734	142671	A3D-2Q	10-58	Active	At NATC in Nov.-Dec. '59 for combined Stability and Control, and Aircraft and Engine Performance Trials. Assigned to VQ-1.
200	11735	142672	A3D-2Q	10-59	1-85	Converted by Douglas as VA-3B for use by CNO. In storage at MASDC from 1974 until retrieved by VAQ-33 team in Sept. '80. Overhauled by NARF Alameda and assigned to VQ-1. Lost at sea 125 miles NNW of Guam on Jan. 23, '85 while operated by VQ-1.
201	11736	142673	A3D-2Q	1-59	Salvgd.	Used as a source of spare parts by NARF Alameda.
202	12024	144626	A3D-2	5-59	11-62	Lost in flight while operated from USS *Forrestal* by VAH-5.
203	12025	144627	A3D-2	6-59	3-67	Lost on March 8, '67 during night mining sortie over North Vietnam (cause unknown) while operated from USS *Kitty Hawk* by VAH-4.
204	12026	144628	A3D-2	6-59	Stored	Converted by NARF Alameda to KA-3B in Aug. '67 and to EKA-3B in Oct. '67. Stored at MASDC since May '78.
205	12027	144629	A3D-2	6-59	10-60	Lost in flight while operated from USS *Independence* by VAH-1.
206	12071	144825	A3D-2P	12-58	Active	At NATC in Sept.-Dec. '59 for carrier suitability trials. Modified by Grumman as NRA-3B with big nose, and assigned to NMC/PMTC since 1963.
207	12072	144826	A3D-2P	6-59	8-69	Lost on Aug. 8, '69 following double engine flame-out while operated by VAP-61 from Da Nang AB on logistics flight.

144831, one of the multi-tone camouflaged RA-3Bs of VAP-61, being serviced in a revetment at Da Nang AB in 1968. *Robert C. Mikesh*

144834, an NRA-3B operated by PMTC for over ten years, has now become the sole TNRA-3B and is assigned to VQ-2 at NS Rota, Spain. *Fred Harl*

144840, an RA-3B of VAP-61, at NAS North Island on November 24, 1968. Now bearing the side number 78, it is operated as an NRA-3B by PMTC. *Duane Kasulka via Norm Taylor*

144843, the RA-3B bailed to Raytheon and used by the U.S. Army for captive tests of the SAM-D missile, at the White Sands Missile Range on November 17, 1975. *U.S. Army via M. J. Kasiuba*

144846, an ERA-3B of VAP-61, was stored at MASDC from July, 1971, until July, 1981. It is now assigned to VAQ-34 as an NRA-3B. *Aerofax, Inc. collection*

Seq. No.	FSN	BuNo	Model	Acc. Date	Disp. Date	Remarks & Disposition
208	12073	144827	A3D-2P	6-59	Active	Converted by NARF Alameda to ERA-3B in June '73 and assigned to VAQ-33.
209	12074	144828	A3D-2P	7-59	6-67	Crashed on June 16, '67 following inflight fire during test flight while operated from NAS Cubi Point by VAP-61.
210	12075	144829	A3D-2P	7-59	6-64	Crashed in North Carolina while operated from NAS Jacksonville by VAP-62.
211	12076	144830	A3D-2P	8-59	5-63	Lost on May 16, '63 while flown from NAS Agana by VAP-61.
212	12077	144831	A3D-2P	9-59	6-76	Salvaged at NARF Alameda.
213	12078	144832	A3D-2P	9-59	Active	Converted by NARF Alameda to ERA-3B in May '73 and assigned to VAQ-33.
214	12079	144833	A3D-2P	10-59	Active	Assigned to PMTC as NRA-3B.
215	12080	144834	A3D-2P	10-59	Active	Converted by NARF Alameda to NRA-3B in July '74 for use at PMTC, and to TNRA-3B in July '85 for use by VQ-2.
216	12081	144835	A3D-2P	10-59	8-67	Failed to return from infrared reconnaissance sortie over North Vietnam on Aug. 25, '67 while operated by VAP-61 from NAS Cubi Point.
217	12082	144836	A3D-2P	11-59	3-60	Lost in flight while operated from NAS Jacksonville by VAP-62.
218	12083	144837	A3D-2P	11-59	4-69	Lost in flight while operated from NAS Jacksonville by VAP-62.
219	12084	144838	A3D-2P	11-59	Active	Retrieved from storage at MASDC and converted by NARF Alameda to ERA-3B in 1981-83 for assignment to VAQ-34.
220	12085	144839	A3D-2P	11-59	11-69	Salvaged at NARF Alameda.
221	12086	144840	A3D-2P	12-59	Active	Converted by NARF Alameda to NRA-3B in Sept. '74 and assigned to PMTC.
222	12087	144841	A3D-2P	12-59	Active	Retrieved from storage at MASDC and converted by NARF Alameda to ERA-3B in 1981-83 for assignment to VAQ-34.
223	12088	144842	A3D-2P	12-59	6-66	Shot down by AAA during photo reconnaissance sortie over North Vietnam on June 13, 1966 while operated by VAP-61 from NAS Cubi Point.
224	12089	144843	A3D-2P	1-60	Active	At White Sands Missile Range, in U.S. Army markings, for captive testing of Patriot (SAM-D) since early '80s; bailed to Raytheon.
225	12090	144844	A3D-2P	1-60	10-67	Shot down by North Vietnamese AAA on Oct. 14, '67 while operated by VAP-61 from NAS Cubi Point.
226	12091	144845	A3D-2P	2-60	9-60	Lost in flight while operated from NAS Jacksonville by VAP-62.
227	12092	144846	A3D-2P	2-60	Active	Retrieved from storage at MASDC and converted by NARF Alameda to ERA-3B in 1982-83 for assignment to VAQ-34.
228	12093	144847	A3D-2P	4-60	1-68	Lost at sea off North Vietnam on Jan. 1, '68 (hit by unknown weapon) while operated by VAP-61 from NAS Cubi Point.
229	12094	144848	A3D-2Q	2-59	12-68	At NATC, as YA3D-2Q, for carrier suitability trials in Oct.-Nov. '59. Damaged beyond economical repair while operated from NS Atsugi by VQ-1.
230	12095	144849	A3D-2Q	3-59	Active	At NATC from June '68 until Jan. '69 for tests of *Seawing* radome. Assigned to VQ-1.

144856, a TA-3B of VAH-123, painted in the high visibility scheme and carrying an Aero 8A practice bomb dispenser beneath its right wing.

144859, one of the five TA-3Bs still operated by VAQ-33 to undertake its responsibility as the A-3 Fleet Replacement Squadron.

144862, a TA-3B of NARU Alameda, the reserve unit which functioned as the A-3 Fleet Replacement Squadron for three years during the mid-seventies.

144867, during its days as the "Phoenix Missile System" testbed. Note APU exhaust below and aft of the national markings on the fuselage side.

146447, modified as an NRA-3B by NARF Alameda in 1972 and since assigned to VAQ-33.

146450, an EA-3B of VQ-1 aboard the USS "Independence" (CVA-62), while participating in "Rolling Thunder" operations during July, 1965.

146452 over the Indian Ocean on November 14, 1974, when this EA-3B of VQ-1 Det 64 took part in "Operation Midlink '74" from the USS "Constellation" (CVA-64).

146457 at Da Nang AB in 1966. More than twenty years later this EA-3B was still assigned to VQ-1.

147650 suffered a nose wheel failure when landing aboard the USS "Coral Sea" during CVA-43's 1965 war cruise. It was lost on October 6, 1966.

147658 photographed at NAS Alameda on October 23, 1968, three days before deployed to the Gulf of Tonkin aboard the USS "Ranger".

Seq. No.	FSN	BuNo	Model	Acc. Date	Disp. Date	Remarks & Disposition
231	12096	144850	A3D-2Q	11-59	1-87	Crashed aboard USS *Nimitz* on Jan. 25 '87.
232	12097	144851	A3D-2Q	11-59	2-70	Lost due to soft cat shot while operated by VQ-2.
233	12098	144852	A3D-2Q	12-59	Active	Assigned to VQ-2.
234	12099	144853	A3D-2Q	11-59	11-59	Missing during ferry flight to VQ-1 at NS Iwakuni.
235	12100	144854	A3D-2Q	11-59	Active	Assigned to VQ-1.
236	12101	144855	A3D-2Q	12-59	9-73	Went down at sea while operated from NAS Agana by VQ-1.
237	12102	144856	A3D-2T	4-59	Active	At NATC for combined Stability and Control, and Aircraft and Engine Performance Trials from Nov. '59 until Feb. '60. Assigned to VAQ-33.
238	12103	144857	A3D-2T	6-59	Active	Converted to VIP-transport configuration by NARF Alameda. Assigned to CFLSW-Det at NAF Washington.
239	12104	144858	A3D-2T	12-59	Active	Assigned to VAQ-33.
240	12105	144859	A3D-2T	1-60	Active	Assigned to VAQ-33.
241	12106	144860	A3D-2T	12-59	5-81	Converted to VIP-transport configuration by NARF Alameda. High-time aircraft and no longer flyable, it was struck off in May '81. Subsequently, it was used by Vought in 1982-83 for Navy-funded stress and fatigue testing.
242	12107	144861	A3D-2T	1-60	8-73	Crashed near Albany, GA, on Aug. 27, '73 while operated by RVAH-3.
243	12108	144862	A3D-2T	2-60	Active	Assigned to VAQ-33.
244	12109	144863	A3D-2T	2-60	7-74	Converted to VIP-transport configuration by NARF Alameda. Crashed while operated from Sigonella, Sicily, by VQ-2.
245	12110	144864	A3D-2T	3-60	2-77	Converted to VIP-transport configuration by NARF Alameda. Salvaged at NARF Alameda.
246	12111	144865	A3D-2T	4-60	Active	Converted to VIP-transport configuration by NARF Alameda. Undergoing SDLM in 1986 and expected to be assigned later to VQ-1.
247	12112	144866	A3D-2T	5-60	Active	Assigned to VAQ-33.
248	12113	144867	A3D-2T	7-60	Active	Bailed to Hughes and then (from Mar. '85) to Grumman.
249	12398	146446	A3D-2P	4-60	Active	Converted by NARF Alameda to ERA-3B in Sept. '72. Assigned to VAQ-33.
250	12399	146447	A3D-2P	4-60	Active	Converted by NARF Alameda to ERA-3B in Sept. '72. Assigned to VAQ-33.
251	12400	146448	A3D-2Q	1-60	Active	Assigned to VQ-2.
252	12401	146449	A3D-2Q	1-60	Stored	Stored at MASDC.
253	12402	146450	A3D-2Q	2-60	8-82	Went down in Indian Ocean on August 4, 1982, while operated by VQ-1 from USS *Ranger*.
254	12403	146451	A3D-2Q	2-60	Active	Assigned to VQ-1.
255	12404	146452	A3D-2Q	4-60	Active	Assigned to VQ-1.
256	12405	146453	A3D-2Q	4-60	Active	Assigned to VQ-2.
257	12406	146454	A3D-2Q	4-60	Active	Assigned to VQ-2.
258	12407	146455	A3D-2Q	4-60	Active	Assigned to VQ-2.
259	12408	146456	A3D-2Q	5-60	1-61	Lost in flight while operated from NS Atsugi by VQ-1.
260	12409	146457	A3D-2Q	5-60	Active	Assigned to VQ-1.
261	12410	146458	A3D-2Q	6-60	11-66	Missing while operated by VQ-2.
262	12411	146459	A3D-2Q	7-60	Active	Assigned to VQ-1.
263	12412	147648	A3D-2	4-60	Active	Converted by NARF Alameda to KA-3B in June '67, to EKA-3B in Oct. '67 and back to KA-3B in Nov. '74. Assigned to VAK-308.

Seq. No.	FSN	BuNo	Model	Acc. Date	Disp. Date	Remarks & Disposition
264	12413	147649	A3D-2	4-60	6-71	Converted by NARF Alameda to KA-3B in June '67 and to EKA-3B in Oct. '67. Lost on June 18, '71 due to control failure while operated from Da Nang AB by VAQ-130.
265	12414	147650	A3D-2	4-60	10-66	Lost on Oct. 6, '66 following mid-air collision with RA-5C while operated by VAH-2 from USS *Enterprise* off Southern California.
266	12415	147651	A3D-2	5-60	10-62	Damaged beyond repair in flight accident while operated off USS *Ranger* by VAH-6.
267	12416	147652	A3D-2	5-60	5-78	Converted by NARF Alameda to KA-3B in June '67 and to EKA-3B in Nov. '67. Salvaged at MASDC.
268	12417	147653	A3D-2	6-60	11-67	Converted to KA-3B by NARF Alameda in June '67. Lost on Nov. 3, '67 when bridle separated prematurely during catapult launch from USS *Constellation* in Gulf of Tonkin while operated by VAH-8 on tanker sortie.
269	12418	147654	A3D-2	6-60	12-63	Lost in flight while operated off USS *Midway* by VAH-8.
270	12419	147655	A3D-2	7-60	Active	Converted by NARF Alameda to EKA-3B in May '67 and to KA-3B in Dec. '74. Assigned to VAK-208.
271	12420	147656	A3D-2	7-60	Active	Converted by NARF Alameda to EKA-3B in May '67 and to KA-3B in Nov. '74. Assigned to VAK-308.
272	12421	147657	A3D-2	8-60	Active	Converted by NARF Alameda to KA-3B in June '67, to EKA-3B in Jan. '68 and back to KA-3B in May '75. Assigned to VAK-208.
273	12422	147658	A3D-2	8-60	Stored	Converted by NARF Alameda to KA-3B in June '67 and to EKA-3B in March '68. Stored at MASDC.
274	12423	147659	A3D-2	9-60	Stored	Converted to EKA-3B by NARF Alameda in June '67. Stored at MASDC.
275	12424	147660	A3D-2	9-60	8-83	Converted by NARF Alameda to KA-3B in June '67, to EKA-3B in Nov. '67 and back to KA-3B in July '74. Struck off in Aug. '83.
276	12425	147661	A3D-2	10-60	9-61	Lost in flight while operated off USS *Ranger* by VAH-6.
277	12426	147662	A3D-2	10-60	11-62	Lost in flight while operated off USS *Ranger* by VAH-6.
278	12427	147663	A3D-2	11-60	Active	Converted by NARF Alameda to EKA-3B in June '67 and to KA-3B in Sept. '74. Assigned to VAK-208.
279	12428	147664	A3D-2	11-60	2-65	Lost on Feb. 24, '65 during tanker sortie (due to fuel transfer failure) while operated by VAH-2 off USS *Coral Sea* in the Gulf of Tonkin.
280	12429	147665	A3D-2	12-60	Active	Converted by NARF Alameda to KA-3B in June '67, to EKA-3B in July '67 and back to KA-3B in Feb. '76. Assigned to VAK-208.
281	12430	147666	A3D-2	12-60	Active	Converted by NARF Alameda to KA-3B in June '67, to EKA-3B in Jan. '68 and back to KA-3B in Oct. '74. Assigned to VAK-308.
282	12431	147667	A3D-2	1-61	Active	Converted by NARF Alameda to KA-3B in June '67, to EKA-3B in July '67 and back to KA-3B in Nov. '74. Assigned to VAK-308.
283	12432	147668	A3D-2	1-61	1-62	Hit mountain in central Luzon, Philippines, on Jan. 13, '62 while operated off USS *Coral Sea* by VAH-2.

147660 on the ramp at NAS Alameda in October, 1982. Shortly thereafter ND 630 was struck off due to corrosion damage.

147663, an EKA-3B of VAW-13 Det 66, over the Atlantic Ocean in March, 1968, when this detachment was working up to deploy aboard the USS "America" (CVA-66).

147665 during a stopover in St. Louis. In early 1987 this EKA-3B was still serving with VAK-208.

147666 breaking away after refueling 147660 during a training sortie southeast of Barstow, California, on November 22, 1981.

147667, seen here in VAK-208 markings but now assigned to VAK-308, was accepted in January, 1961, and is thus the youngest "Skywarrior" still in service.

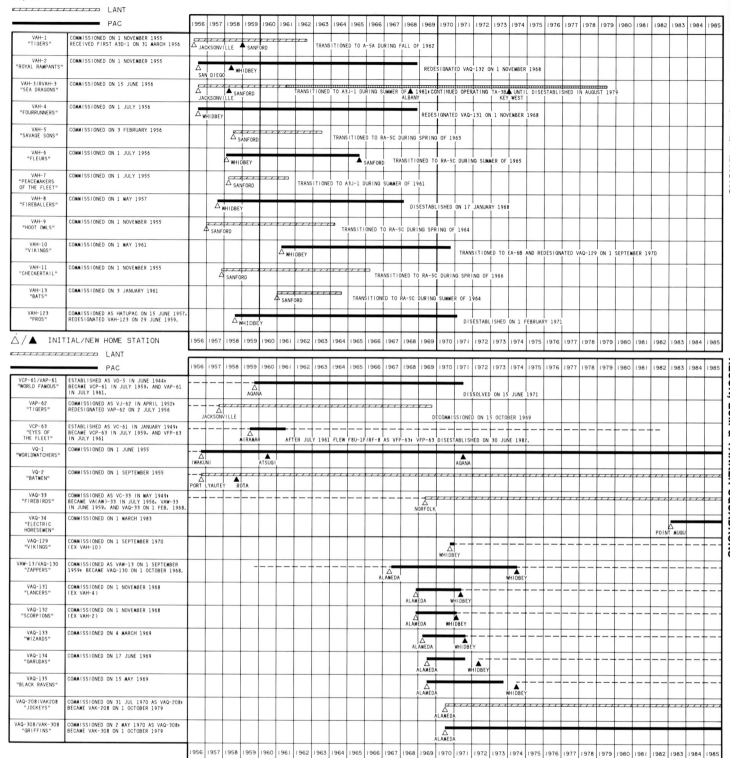

Bibliography

Books:

Air War Over Southeast Asia, Volumes 1, 2 and 3, Lou Drendel, Squadron/Signal Publications, Carrollton, Texas, 1982, 1983 and 1984.
Aircraft Carriers — A Graphic History of Carrier Aviation and Its Influence on World Events, Norman Polmar, Doubleday & Company, Inc., Garden City, New York, 1969.
American Combat Planes, Ray Wagner, Doubleday & Company, Inc., Garden City, New York, 1960 and 1968.
Bombers of the West, Bill Gunston, Ian Allan Ltd., Shepperton, England, 1973.
Combat Aircraft of the World, Edited and compiled by John W. R. Taylor, G. P. Putnam's Sons, New York, New York, 1969.
Ed Heinemann — Combat Aircraft Designer, Edward H. Heinemann and Rosario Rausa, Naval Institute Press, Annapolis, Maryland, 1980.
The History of U.S. Naval Air Power, Editor: Robert L. Lawson, Temple Press/Aerospace Publishing Ltd., London, England, 1985.
The Naval Air War In Vietnam, Peter B. Mersky & Norman Polmar, The Nautical and Aviation Publishing Company of America, Annapolis, Maryland, 1981.
United States Navy Aircraft Since 1911, Gordon Swanborough & Peter M. Bowers, Putnam & Company Ltd., London, England, 1968.
U.S. Aircraft Carriers — An Illustrated Design History, Norman Friedman, Naval Institute Press, Annapolis, Maryland, 1983.
USN Aircraft Carrier Air Units, Volume 2, 1957 — 1963, Duane Kasulka, Squadron/Signal Publications, Carrollton, Texas, 1985.
World Electronic Warfare Aircraft, Martin Streetly, Jane's Publishing Company Ltd., London, England, 1983.

Magazines:

Air Fan, Paris, France, various issues.
Air International, London, England, various issues.
Air Pictorial, London, England, various issues.
Airpower, Granada Hills, CA, various issues.
Air World, Vol. 9, No. 11, Tokyo, Japan.
Approach, Norfolk, VA, various issues.
Aviation Week & Space Technology, New York, NY, various issues.
Flight International, London, England, various issues.
Naval Aviation News, Washington, D.C., various issues.
RAF Flying Review, London, England, various issues.
The Hook, Bonita, CA, various issues.
Warplane, London, England, various issues.
Wings, Granada Hills, CA, various issues.

Douglas Reports, Manuals, and other Documents:

Bar Chart, A3D Model History. Undated.
Douglas Technical Drafting, Technical Support Drawings A3D-2P. September 15, 1959.
Douglas Technical Drafting, Technical Support Drawings A3D-2Q. September 15, 1959.
Douglas Technical Drafting, Technical Support Drawings A3D-2T. September 15, 1959.
Report No. ES 17713, A3D Skywarrior Carrier-based Attack Bomber. November 22, 1954.

U.S. Navy Reports, Manuals, and other Documents:

Board of Inspection and Survey, Report of Service Acceptance Trials on Model A3D-2 Aircraft. March 30, 1959.
Board of Inspection and Survey, Report of Service Acceptance Trials on Model A3D-2T Aircraft. September 20, 1960.
NASWF Project TED No. BIS 21178, Service Acceptance Trials, A3D-1 Aircraft, Special Weapons Phase. January 29, 1957.
NATC Project No. PTR AC-317.1, Navy Preliminary Evaluation of Model A3D-1 Airplane, Report No. 2, Special Report, Addendum to. December 1, 1954.
NATC Project No. PTR AC-317.6, Fuel Consumption Tests of the Model A3D-1 Airplane, Report No. 1, Final Report. August 14, 1957.
NATC Project TED No. BIS 21213, Carrier Suitability Trials of the Model A3D-1 Airplane, Report No. 2; Arrested Landing Tests of the A3D-1 Airplane with Special Stores and Evaluation of Modified Forward Fuel Tank Installation. August 29, 1957.
NATC Project No. PTR SI-4282, Shipboard Carrier Suitability Trials of Model A3D-2 Airplane, Special Report of. May 21, 1958.
NATC Project TED No. BIS 21213, Combined Stability and Control and Aircraft and Engine Performance Trials of the Model A3D-2 Airplane, Report No. 1, Final Report. September 4, 1958.
NATC Project No. BIS 21213, Service Suitability Trials of Model A3D-2 Airplane, Report No. 1, Interim Report. September 19, 1958.
NATC Project No. PTR SI-4282, Carrier Suitability Trials of Model A3D-2 Airplane, Report No. 3, Supplement to Final Report. April 22, 1959.
NATC Project No. PTR AC-3005.6, Fuel Consumption Test of the Model A3D-2 Airplane, Report No. 1, Final Report. May 4, 1959.
NATC Project TED PTR AE-6206, Model A3D-2 Air Refueling System, Tanker/Receiver, Service Suitability Evaluation, Report No. 1, Final Report. August 21, 1959.
NATC Report No. FT-9R-66, Fifth Interim Report, Flight Evaluation of Approach Power Compensator. February 7, 1966.
NATC Report No. FT-27R-66, Final Report, Investigation of Feasibility of Performance Demonstration of Landing Approach Speed, including A-3B Phase. March 16, 1966.
NATC Project No. BIS 21229, Combined Stability and Control and Aircraft and Engine Performance Trials of the Model A3D-2P Airplane, Report No. 1, Final Report. November 27, 1959.
NATC Project No. BIS 21229, Carrier Suitability Trials of the Model A3D-2P Airplane, Report No. 1, Final Report. March 2, 1960.
NATC Project No. BIS 21230, Carrier Suitability Trials of the Model A3D-2Q Airplane, Report No. 1, Final Report. March 10, 1960.
NATC Project No. BIS 21230, Combined Stability and Control and Aircraft and Engine Performance Trials of the Model A3D-2Q Airplane, Report No. 1, Final Report. March 31, 1960.
NATC Project No. FT-010R-69, Final Report, Preliminary Evaluation of the EA-3B Airplane with Modified Radome. June 16, 1969.
NATC Project No. BIS 21239, Combined Stability and Control and Aircraft and Engine Performance Trials of the Model A3D-2T Airplane, Report No. 1, Final Report. June 15, 1960.
NAVAIR 00-80P-1, United States Naval Aviation, 1910-1980. 1981.
NAVAIR 01-40AT-75, Conventional Weapons Loading Manual - Navy Model A-3B, RA-3B and TA-3B Aircraft. February 1, 1967.
NAVAIR 01-40ATA-1, NATOPS Flight Manual - Navy Model A-3A/B Aircraft (Including Tanker Configurations). September 1, 1966.
NAVAIR 01-40-ATA-2-2, Handbook Maintenance Instructions - Navy Models A-3A, A-3B, KA-3B, and EKA-3B Aircraft; Section II, Airframe, Surface Controls, Landing Gear, Arresting Gear, and JATO. April 15, 1969.
NAVAIR 01-40ATB-2s, Supplemental Technical Manual, Maintenance Instructions - Navy Model ERA-3B Aircraft; Aircraft, Navigation and Communications, and Electronic Warfare Systems. July 1, 1974.
NAVAIR 01-40ATC-2-1, Handbook Maintenance Instructions - Navy Model EA-3B Aircraft; Section I, General Information. October 15, 1971.
NAVAIR 01-40ATD-2-2, Handbook Maintenance Instructions - Navy Model A3D-2T Aircraft; Section II, Airframe, Surface Controls, Landing Gear, Arresting Gear, and JATO. December 1, 1965.
NAVAIR 01-40ATE-1, NATOPS Flight Manual - Navy Model EKA-3B and KA-3B Aircraft. January 1, 1976.
NAVAIR, Standard Aircraft Characteristics
  A3D-1, April 1, 1952, April 15, 1957, and September 1, 1958.
  A3D-1P, September 11, 1953, and July 1, 1954.
  A3D-2 (Uncambered Wing), April 15, 1961.
  A3D-2 (Cambered Wing), September 1, 1957, and April 15, 1961.
  A3D-2 (Tanker), March 3, 1958.
  A3D-2P, November 25, 1955, and April 15, 1961.
  A3D-2Q, December 9, 1960, and April 15, 1961.
  A3D-2T, April 14, 1958.
  D-970 Missileer, February 19, 1960.
NAVAIR, A3D History, August 1948-March 1957.
NAVWEPS 01-40ATC-4-5, Illustrated Parts Breakdown - Navy Models A3D-2Q and A3D-2T Aircraft; Volume V, Electrical Power and Electronics. February 15, 1962.
NAVWEPS 01-40ATA-4-7, Illustrated Parts Breakdown - Navy Models A-3A, A-3B, KA-3B and EKA-3B Aircraft, Volume VII Combat Equipment. April 1, 1977.

Sundry Reports:

Probe and Drogue Refueling Systems, A General Overview. Flight Refueling, Inc. Columbia, MD. Undated.

Index

Adak, AK: 45, 62
Aerospace Recovery Facility (ARF): 71
Air bases, stations, and facilities:
  Aberdeen Bombing Range: 21
  Andrews AFB: 48
  Buckley ANGB: 129
  DaNang AB: 47, 54, 57, 59, 64, 66, 70-71, 85, 129-130, 132-133
  Davis Monthan AFB: 26, 37, 39, 70
  Edwards AFB: 3, 16-17, 19-24, 31, 46, 121-123
  Elmendorf AFB: 57, 70, 128
  Hickam AFB: 24, 72
  Howard AFB: 46
  King Salmon AFB: 45
  Kirtland AFB: 23, 72
  Kwajalein Army Test Site: 33, 72
  MCAF Iwakuni/NS Iwakuni: 46, 70, 122, 132
  Misawa AB: 35, 45, 48
  Muroc AFB: 12
  NADC Johnsville: 31-32, 35, 71, 80, 122
  NAF El Centro: 71, 121
  NAF Washington: 35-37, 48, 71, 125, 127
  NAS Agana: 32, 35, 37, 46, 48, 64, 66, 70, 116, 130, 132
  NAS Alameda/NARF Alameda: 21, 26-27, 30, 32-34, 36-37, 40, 43, 46, 47, 48, 53-56, 58-60, 62-63, 65-70, 72-73, 81, 83, 86-88, 95, 107, 112, 121-133
  NAS Albany: 57
  NAS Anacostia: 9
  NAF Atsugi/NAS Atsugi/NS Atsugi: 45, 69-70, 86, 130, 132
  NAS Barbers Point: 44-45, 60, 68, 82-83
  NAS Brooklin: 45
  NAS China Lake/NWS China Lake: 26, 48, 54, 73, 125-126
  NAS Cubi Point: 32, 47, 54, 57, 85-87, 125, 130
  NAS Floyd Bennett: 61
  NAS Glenview: 58
  NAS Jacksonville/O&R Jacksonville: 24, 28-29, 41-42, 46, 55-56, 58, 64, 86, 123, 129-130
  NAS Key West: 56-57, 59, 63, 71, 107, 123
  NAS Lakehurst/NATF Lakehurst: 25, 73, 122, 124, 127
  NAS Miramar: 46, 73, 82, 129
  NAS Moffett Field: 12, 24, 44, 49, 58-59, 62
  NAS Naha: 86
  NAS North Island/NARF North Island: 56, 62, 71, 90, 130
  NAS Norfolk/NARF Norfolk/O&R Norfolk: 12, 24, 36-37, 46, 58, 64-65, 67, 122-123, 125-128
  NAS Oceana: 40, 65
  NAS Patuxent River/NATC Patuxent River: 12, 21-24, 27, 29, 31, 35, 45, 55-56, 59, 71-72, 80, 84, 122, 124, 127
  NAS Pensacola: 113
  NAS Point Mugu/PMTC Point Mugu: 21, 26, 34, 37-38, 48, 54, 65, 73, 112-113, 121-122, 124-125, 129
  NAS Rota/NS Rota: 35, 46-48, 71, 87, 130
  NAS Sanford: 43, 45, 55, 57-62, 88, 123, 125-130
  NAS Whidbey Island: 45, 51, 56-57, 59-62, 66-69, 72, 86, 122-126, 128
  NAS Willow Grove: 37
  NASWF Albuquerque: 24
  NS Sangley Point: 70
  NWL Dahlgreen: 123
  Offutt AFB: 63
  Pacific Proving Grounds: 24
  Port Lyautey: 12, 23, 46, 71, 107, 121-122
  Scott AFB: 113
  Shemya AFB: 35
  Wendover AFB: 46
  White Sands Missile Range: 46, 130
Air Groups and Air Wings:
  ATG-1: 51
  CVG-1/CVW-1: 35, 42, 50, 52, 61, 123, 127
  CVG-2/CVW-2: 51-54, 68
  CVG-3/CVW-3: 50
  CVG-5/CVW-5: 51-53
  CVG-6/CVW-6: 50, 52-53, 59, 67, 69
  CVG-7/CVW-7: 42, 50, 52, 55
  CVG-8/CVW-8: 50, 52-53, 59, 61, 67
  CVG-9/CVW-9: 51-54, 67, 88
  CVG-10: 50, 107
  CVG-11/CVW-11: 51-54, 62, 68
  CVG-14/CVW-14: 44, 51-54, 68
  CVG-15/CVW-15: 51-54, 85
  CVG-16/CVW-16: 51-54
  CVG-17/CVW-17: 42, 50, 52
  CVG-19/CVW-19: 51-53, 82
  CVG-21/CVW-21: 51-54, 60, 66, 69, 79, 82
  CVWR-20: 48, 63, 128
  CVWR-30: 48, 63
Aircraft engines:
  Aerojet KS-4500 JATO: 25, 46, 57, 119
  Allison J33: 11
  Allison T40: 12, 14
  Allison T44: 15
  Allison T56: 29
  Garrett ALF-502: 48-49
  General Electric J79: 30, 37
  General Electric TF34: 48-49
  Ghome et Rhône 14R: 10
  Marquardt (thrust reverser on J57): 21
  Pratt & Whitney J57: 19-23, 25, 28, 49, 106, 118-120
  Pratt & Whitney J75: 23, 26, 28-29
  Pratt & Whitney JT3: 19
  Pratt & Whitney JT3D: 30
  Pratt & Whitney JT4A: 28-29
  Pratt & Whitney JT8D: 30
  Pratt & Whitney R-1535: 9
  Pratt & Whitney R-2800: 9, 11
  Pratt & Whitney R-4360: 9
  Pratt & Whitney TF30: 30
  Rolls-Royce Nene: 10
  Westinghouse J34: 19
  Westinghouse J40: 3, 18-21, 23, 25
  Wright J67: 28
  Wright P-2: 9
Aircraft:
  Breguet 810: 9
  Boeing 707: 30, 36
  Boeing 737: 48
  Boeing B-29: 7-8, 12-13
  Boeing B-47: 13, 15, 19
  Boeing B-50/KB-50: 27, 32
  Boeing B-52: 49
  Boeing KC-135: 30, 47
  C.A.O. 600: 9
  Consolidated XB-46: 13, 15
  Convair B-58: 45
  Curtiss XVA(H1): 15, 17
  de Havilland C-8 (QSRA): 48
  de Havilland Mosquito/Sea Mosquito: 9-10
  de Havilland Sea Hornet: 9-10
  de Havilland Sea Vampire: 10
  Dewoitine D.750: 8-9
  Douglas A-1 (AD): 17, 27, 34, 45, 49, 65-66, 85
  Douglas A-4 (A4D) & TA-4: 21, 39-40, 47, 49, 85

  Douglas B/RB-66: 19-21, 71
  Douglas C-124: 17
  Douglas D-558: 15, 17
  Douglas DC-6: 17
  Douglas DC-8: 21, 28-29, 36
  Douglas DC-9: 30
  Douglas *Missileer*: 30, 38
  Douglas F3D: 15, 17-18, 30, 43, 56, 62
  Douglas F6D: 30, 38
  Douglas R4D: 10-11, 57
  Douglas RB-26: 19
  Douglas SBD: 49
  Douglas TBD-1: 9
  Douglas XA2D-1: 17
  Douglas XB-43: 15-16
  Douglas XF4D-1: 17, 19
  Douglas XT2D-1: 9
  Douglas XTB2D-1: 8
  Douglas X-3: 15
  Grumman A-6: 47-48, 62
  Grumman EA-6: 27, 47-48, 53, 66-69
  Grumman F7F-1: 9
  Grumman F9F: 27, 43, 57, 61, 86
  Grumman F-14: 36, 39, 48, 112
  Grumman F-111B: 38, 113, 117
  Grumman TBM-3: 65
  Grumman XF5F-1: 9
  Grumman XF10F-1: 9
  Grumman XTB2F-1: 9
  Grumman XTSF-1: 9
  Gulfstream American C-20: 36
  Levasseur PL.7: 9
  Lockheed EC-121 (WV-2): 70-71
  Lockheed EP-3: 70-71
  Lockheed F-104: 47
  Lockheed P2V: 12, 17, 35, 43, 56, 58-59, 62
  Lockheed S-3: 63
  Lockheed T-33 (TV-2): 43, 48, 56
  Lockheed U-2: 42, 44, 113
  Lockheed XJ0-3: 9
  LTV F/RF (F8U): 29, 45-47, 64, 85
  Martin P4M: 34, 46, 70-71
  Martin P6M: 28-29
  Martin RB-57: 19
  Martin XB-48: 13, 15
  Martin XB-51: 19
  McDonnell F-3 (F3H): 19, 45
  McDonnell F-4 (F4H): 45, 47, 55, 60, 88, 128
  McDonnell RF-101: 45, 47
  McDonnell Douglas DC-10: 47
  McDonnell Douglas F-18: 72
  Naval Aircraft Factory XTN-1: 9
  Nord 1500: 8, 10
  North American A/RA-5 (A3J): 28, 35, 41, 43, 45, 47, 55, 57-62, 107
  North American AJ: 9, 11-13, 17, 27, 42-43, 58-62, 64, 88
  North American B-25: 7, 9
  North American B-45: 19
  North American CT-39: 71
  North American F-86: 24
  North American F-100: 48
  North American FJ: 45
  North American PBJ-1: 7, 9
  North American X-15: 46
  Potez 56E: 6, 9
  Republic F-84F: 24
  Short *Sturgeon*: 10
  S.N.C.A.C. N.C. 1070: 10
  S.N.C.A.C. N.C. 1071: 10
  Tupolev Tu-142: 87
  Vega V-141: 9
  Vickers *Valiant*: 19
  Vought A-7: 65
  Vought F4U: 49
  Albuquerque, NM: 45, 116
  Ames Laboratory/Ames Research Center: 21, 48-49
  Ashworth, Frederic L., Cdr., USN: 8
  Ballhaus, Dr. William F.: 15
  Baum, Russell H., Lt. USN: 24
  Berk, Paul D.: 29
  Blackburn, J. T., Capt., USN: 43
  Bradley Air Museum: 72, 84, 126-127
  Brown, Eric LCdr., RN: 10
  Butts, Robert L., TSgt., USMC: 24
  California Institute of Technology: 19
  Cam Ranh Bay: 87
  Cape Canaveral/Cape Kennedy: 46
  Chaffee, Robert B.: 42
  Charleston, SC: 45-46, 127
  Clark, Thurston B., LCdr., USN: 7, 9
  Coanda flaps: 48
  Combs, Thomas S., ViceAdm., USN: 41
  Commander, Fleet Logistics Support Wing (COMFLELOGSUPPWING or CFLSW): 54, 71, 132
  Cox, Dale W., Cdr., USN: 24
  Dangler, Mike, Lt., USN: 87
  Davidson, James, LCdr., USN: 10
  Davies, T. D., Cdr., USN: 12
  Davis, Bill: 23-24
  Dayton, OH: 23, 25
  Decker, Edward A., LCdr., USN: 27
  de Lorenzi, R. M., Cdr., USN: 85
  de L'Orza, L. V., Marine Nationale: 6, 9
  Devlin, Leo J.: 15, 18
  Donnelly, W. N. Cdr., USN: 85
  Driggs, Ivan H.: 14
  Edwards, Terry W., Lt(jg) USN: 85
  El Segundo, CA: 1, 13-15, 17, 20-21, 31, 34, 41, 49
  Eniwetok: 24, 33, 71
  Fleet Electronic Warfare Support Group (FEWSG): 37, 39, 48, 65, 71, 124, 126
  Flight Refueling, Inc: 27
  Frossard, Cdr., USN: 24
  Glenn, John H., Jr.: 46
  Gordon, Richard F., Lt., USN: 45
  Guam: 36, 130
  Guantanamo: 9
  Gulf of Tonkin: 28, 37, 46-47, 52, 56, 60-61, 66-68, 88, 125, 127-130, 132-133
  Guggenheim Aeronautical Laboratory: 19
  Hall, G. Warren: 48
  Halsey, William F., Adm. USN: 7
  Harris, J. H., Cdr., USN: 85
  HATULANT: 43
  HATUPAC: 43, 62
  Hayward, John T., Cpt., USN: 12, 17
  Heinemann, Edward H.: 4, 9, 13-16, 18-20, 24, 27, 56, 117
  Highball: 9

  Hiroshima: 7, 8
  Ho Chi Minh Trail: 47, 64
  Hume, K. E., LCdr., USN: 85
  Ile Bach Long Vi: 85
  Ile Nightingale: 85
  Jansen, George R.: 3, 20
  Kane, Ed, Lt., USN: 87
  Kent, Walter: 3, 20
  Kien Giang River: 54
  Laos, 27, 48
  Laviano, Mike, Lt., USN: 87
  Libya: 54, 71
  Lippitt, David E., LT1, USN:85
  Litchfield Park: 23, 29, 121-123
  Little America: 10-11
  Long Beach, CA: 17, 111
  Los Angeles, CA: 20, 23-24
  Los Angeles IAP: 24, 31
  Lyon, Lt., USN: 9
  MASDC: see Military Aircraft Storage & Disposition Center
  Masters, Randy, Lt., USN: 87
  Mayport, FL: 43, 123
  Military Aircraft Storage & Disposition Center (MASDC): 26-27, 37-38, 40, 49, 66, 70, 121-133
  Mitchell, Edgar D: 42, 56
  Mitscher, Marc A., Vice Adm., USN: 8-10
  Moffett, William A. Rear Adm., USN: 9
  Mongilardi, P., Cdr., USN: 85
  Moreland, "Speed", Cdr., USN: 44
  Morgenson, D. L., Lt(jg), USN: 85
  Nagasaki: 7-8
  Napa, CA: 128
  NARU Alameda: 43, 72, 131
  National Advisory Committee for Aeronautics (NACA): 21
  National Aeronautics and Space Administration (NASA): 48-49
  National Atomic Museum, Albuquerque: 116
  National Parachute Test Range (NPTR): 71, 128
  Naval Airborne Project Operations Group (NAPOG): 33, 72
  Naval Air Development Center (NADC): 32, 71, 122 (see also NADC Johnsville)
  Naval Aviation Museum, Pensacola: 38, 113, 122
  Naval Air Special Weapons Facility (NASWF): 24, 72, 122
  Naval Air Technical Training Center (NTTC): 19, 22, 121
  Naval Air Test Center (NATC): 23, 72, 122, 126-128, 130-132 (see also NAS Patuxent River)
  Naval Air Test Facility (NATF): 73 (see also NAS Lakehurst)
  Naval Missile Center: 21, 33, 73, 113, 121-122, 125, 129-130 (see also NAS Point Mugu)
  Naval Ordnance Test Station (NOTS): 73 (see also NAS China Lake)
  Naval Procurement Office (NAVPRO) Dallas: 128
  Naval Weapons Center (NWC): 73, 128
  Naval Weapons Evaluation Facility (NWEF): 72, 121
  New York: 22-23
  Nichols, Harry A.: 15, 18
  Norfolk, VA: 12, 46, 67
  North Africa: 7
  North Vietnam: 27, 46, 48, 54, 64
  Northrop, John K.: 9
  North American PBJ-1: 7, 9

  Operation *Highjump*: 11
  Operation *Midlink 74*: 132
  Operation *Olympic-Majestic*: 7
  Operation *Rimpac*: 40, 48
  Pacific Missile Test Center (PMTC): 73, 113, 124-125, 129-131 (see also NAS Point Mugu)
  Panama: 12, 46
  Pan American Airways: 36
  Pangkalan Brandon: 7
  Parker, Charles, LCdr., USN: 18
  Parsons, William, Capt./Rear Adm., USN: 8, 12
  Pima County Air Museum: 34, 122
  Pirie, R. B., Vice Adm., USN: 18, 36
  Polaris: 28, 45
  Pride, Alfred M., LCdr., USN: 9
  Project *AROWA*: 24
  Project *Autec*: 46
  Project *LANA*: 45, 55, 57, 60, 128
  Project *PRESS*: 32-33, 72, 127
  Project *Redwing*: 7
  Project *Stormfury*: 45-46
  Radford, Arthur W., Vice Adm., USN: 12
  Ramage, J. D., Rear Adm., USN: 43
  Rees, William L., Vice Adm., USN: 41
  Richards, L. W., LCdr., USN: 85
  Robinson, T., Cdr., USN: 12
  Root, I. Eugene: 15
  Rousselton, Congressman: 18
  Sallada, Harold B., Rear Adm., USN: 10-11, 13
  Salyer, H. L., Cdr., USN: 56
  San Francisco, CA: 12, 45
  Santa Monica, CA: 13, 17
  Seawing: 131
  17th Bombardment Group: 7
Ships:
  Béarn: 6, 9
  HMS *Ark Royal*: 10
  HMS *Hermes*: 10 HMS *Illustrious*: 7
  HMS *Indefatigable*: 9
  HMS *Ocean*: 10
  Joffre: 8
  Painleve: 8
  USS *America* (CVA/CV-66): 35, 50, 52-54, 66-67, 69, 128, 133
  USS *Bon Homme Richard* (CVA-31): 24, 42, 44, 51-53, 56, 66, 83, 124, 126
  USS *Constellation* (CVA/CV-64): 44, 46, 51-54, 60-61, 67-68, 123, 132-133
  USS *Coral Sea* (CVB/CVA/CV-43): 9, 12, 42, 45, 48, 51-54, 56, 69, 85, 111, 125, 127-129, 132-133
  USS *Enterprise* (CVAN/CVN-65): 52-54, 66-67, 88, 123-124, 130, 133
  USS *Essex* (CV/CVA-9): 14, 16, 24
  USS *Forrestal* (CVA-59): 24, 41-42, 50, 52, 55, 58-59, 61, 69, 72, 88, 107, 124-125, 127-128, 130
  USS *George Washington* (SSBN-598): 45
  USS *Hancock* (CVA-19): 42-43, 49, 51-54, 57, 60, 66, 69, 82, 86, 127, 129
  USS *Hornet* (CV-8): 7, 9
  USS *Independence* (CVA-62): 27, 42, 50, 52, 55, 57, 59, 61, 72, 125, 127
  USS *Intrepid* (CVA/CVS-11): 2, 20, 121
  USS *John F. Kennedy* (CVA-67): 35, 50, 65, 67, 69, 106, 127

  USS *Kitty Hawk* (CVA/CV-63): 47-48, 51-54, 61-62, 67 60, 87, 124, 129-130
  USS *Langley* (CV-1): 9
  USS *Lexington* (CV-2): 7, 9
  USS *Lexington* (CVA-16): 51
  USS *Maddox* (DD-731): 46
  USS *Midway* (CVB/CVA-41): 8, 12, 14, 16, 41, 44, 45, 51-53, 59-60, 86, 128, 133
  USS *Nimitz* (CVN-68): 35
  USS *Oriskany* (CVA-34): 47-48, 51-54, 59, 86, 124-125, 128-129
  USS *Philippine Sea* (CV-47): 11-10
  USS *Ranger* (CVA-61): 19, 44, 48, 51-54, 56, 58-59, 66, 68, 86, 116, 126-128, 132-133
  USS *Franklin D. Roosevelt* (CVB/CVA/CV-42): 10, 12, 42, 50, 52, 57, 61, 69, 123-124, 126-128
  USS *Saratoga* (CV-3): 7
  USS *Saratoga* (CVA/CV-60): 42, 44-45, 50, 60, 66, 123-126
  USS *Shangri-La* (CVA/CVS-38): 7, 9, 41, 44, 51-53, 66, 86, 125
  USS *Ticonderoga* (CVA-14): 46, 51-53, 56-57, 123
  USS *United States* (CVB/CVA-58): 8, 14-16
  USS *Yorktown* (CVS-10): 127
  Smith, R. G.: 15
  Somerville, Sir James, Adm., RN: 7
Squadrons:
  RVAH-1: 46, 55
  RVAH-3: 35, 56-57, 74, 107, 119, 132
  RVAH-5: 58
  RVAH-6: 59
  RVAH-7: 59
  RVAH-9: 60
  RVAH-11: 61
  RVAH-13: 61-62
  VA-63: 45
  VA-65: 45
  VA-153: 85
  VA-155: 85
  VA(AW)-33: 65
  VA(AW)-35: 65
  VAH-1: 41-45, 50, 55, 59, 77, 122-123, 125, 127, 129-130
  VAH-2: 26, 28, 41-45, 51-52, 54, 56, 61, 67, 85, 86, 111, 125
  VAH-3: 26, 35, 41-43, 46, 50, 56-57, 122-123
  VAH-4: 41-47, 51-52, 54, 57, 61, 67, 81, 86, 88, 123-127, 129-130, 132
  VAH-5: 42-43, 46, 50, 88, 107, 124-125, 130
  VAH-6: 42-43, 46, 50-51, 58-59, 86, 116, 124, 126-128, 133
  VAH-7: 42-45, 59, 123
  VAH-8: 42-45, 47, 51-52, 54, 59-60, 86, 126, 128, 133
  VAH-9: 42-46, 50, 60, 123-126, 128
  VAH-10: 42-43, 45-47, 50-52, 54, 60-61, 66, 79, 82, 123, 126-127
  VAH-11: 42-43, 46, 50, 61, 123-129
  VAH-13: 42-43, 45-46, 51, 61-62, 125-126
  VAH-123: 26, 42-43, 62, 66, 83, 122-125, 131
  VAK-208: 37, 48, 54, 63, 79, 82, 128, 130
  VAK-308: 37, 48, 54, 63, 79, 83, 85, 128, 130, 132-133
  VAP-61: 31-32, 39, 41, 47-48, 54, 64, 70, 81, 116, 130-131
  VAP-62: 31, 41-42, 46-48, 64, 86, 131
  VAQ-33: 35, 39-40, 43, 48, 54, 62, 65, 71-72, 82, 123-126, 129-132
  VAQ-34: 39-40, 48, 54, 65, 70-71, 84, 125, 130-131
  VAQ-129: 50, 53, 68
  VAQ-130: 43, 50-51, 53-54, 59, 62, 66-67, 69, 74, 82, 86, 127-128, 132-133
  VAQ-131: 50, 53, 57, 67, 76, 124, 127
  VAQ-132: 53-54, 56, 67, 123, 126, 128
  VAQ-133: 53, 68
  VAQ-134: 53, 68
  VAQ-135: 48, 50, 53-54, 69, 86-87, 106, 126-129
  VAQ-208: 48, 63, 86, 87
  VAQ-308: 48, 63, 125, 129
  VAW-13: 53, 66-67, 82-83, 132
  VAW-33: 65
  VC-1: 39-40
  VC-5: 9, 11-12, 58-59
  VC-6: 12, 59
  VC-9: 60
  VC-11: 61
  VC-33: 65
  VC-61: 64, 69
  VCP-61: 46, 64
  VCP-63: 31, 46, 69
  VD-5: 64
  VF-64: 45
  VF-96: 88
  VF-101: 60, 128
  VF-154: 85
  VF-211: 45
  VFP-63: 46, 69, 85
  VJ-62: 11, 64
  VP-16: 55
  VP-29: 56
  VP-34: 56
  VP-57: 57
  VQ-1: 32, 34-37, 41, 46-48, 54, 70-71, 80, 87, 106, 122, 130-132
  VQ-2: 23, 34-35, 37, 41, 46-48, 54, 70-71, 87, 106, 121-122, 128, 130-132
  VR-1: 36, 71
  VX-5: 41
  Starkey, R. C., LCdr., USN:12
  Stephens, Paul F., Cdr., USN: 41
  Stewart, Tom, LCdr., USN: 44
  St. Louis: 42, 45, 63, 133
  Sullivan, John L. Actg Sec Navy: 8
  Sundberg, John L., Cdr., USN: 85
  Systems and equipment: 18-19, 21, 24-28, 31, 34-35, 37-40, 43, 45, 47, 62, 85, 92-93, 95-98, 101, 107-110, 113, 116-117, 119-120, 131
  Task Force 38: 7
  Thompson, James L. ABE1, USN: 85
  Toft, Richard J. Lt(jg), USN: 85
  Upper Surface Blowing (USB): 48-49
  Van Every, Kermit E.: 15
  Vaughan, Robert R., LCdr., USN: 85
  Wangeman, C. E., Jr., Lt., USN: 85
  Warde, William LCdr., USN: 85
  Weapons: 8, 12-13, 18, 21, 24-26, 30-31, 34, 36-38, 45, 62, 72, 85, 112, 114-117, 122-123, 131
  Wheatley, J. P. Cdr., USN: 12
  Willis, Don J., Lt(jg), USN: 85
  Yokosuka Navy Yard: 7
  Young, Bobbie R., Lt(jg), USN: 45